MW00332186

Limits of the Numerical

．．．

Limits of the Numerical

．．．

THE ABUSES AND USES
OF QUANTIFICATION

Edited by
Christopher Newfield,
Anna Alexandrova,
and Stephen John

THE UNIVERSITY OF CHICAGO PRESS
CHICAGO AND LONDON

The University of Chicago Press, Chicago 60637
The University of Chicago Press, Ltd., London
© 2022 by The University of Chicago
All rights reserved. No part of this book may be used or reproduced
in any manner whatsoever without written permission, except in
the case of brief quotations in critical articles and reviews.
For more information, contact the University of Chicago Press,
1427 East 60th Street, Chicago, IL 60637.
Published 2022
Printed in the United States of America

31 30 29 28 27 26 25 24 23 22   1 2 3 4 5

ISBN-13: 978-0-226-81713-2 (cloth)
ISBN-13: 978-0-226-81715-6 (paper)
ISBN-13: 978-0-226-81716-3 (e-book)
DOI: https://doi.org/10.7208/chicago/9780226817163.001.0001

Library of Congress Cataloging-in-Publication Data

Names: Newfield, Christopher, editor. | Alexandrova, Anna,
1977– editor. | John, Stephen, editor.
Title: Limits of the numerical : the abuses and uses of quantification /
edited by Christopher Newfield, Anna Alexandrova, and Stephen John.
Description: Chicago ; London : The University of Chicago Press,
2022. | Includes bibliographical references and index.
Identifiers: LCCN 2021050921 | ISBN 9780226817132 (cloth) |
ISBN 9780226817156 (paperback) | ISBN 9780226817163 (ebook)
Subjects: LCSH: Quantitative research. | Quantitative research—Case
studies. | Quantitative research—Social aspects. | Education, Higher—
Research—Case studies. | Health—Research—Case studies. |
Climatology—Research—Case studies.
Classification: LCC Q180.55.Q36 L56 2022 | DDC 001.4/2—dc23/
eng/20211204
LC record available at https://lccn.loc.gov/2021050921

The University of Chicago Press gratefully acknowledges the generous
support of the Independent Social Research Foundation toward the
publication of this book.

♾ This paper meets the requirements of ANSI/NISO Z39.48-1992
(Permanence of Paper).

Contents

The Changing Fates of the Numerical

Christopher Newfield, Anna Alexandrova, and Stephen John

Both private and commercial aircraft have a variety of navigational tools they can use. One is very high frequency omnidirectional range and distance measuring equipment (VOR/DME), which allows users to combine measures of their bearing and their slant distance from an object like an airport runway. It uses highly detailed quantitative information to offer precise measures of the kind that allow pilots to execute safe landings in bad weather without good visual contact with the ground. As such, VOR/ DME offers a classic example of the powers of the numerical as they improve daily life by building quantification into systems.

Not long after midnight on August 6, 1997, Korean Air Lines (KAL) flight 801 was using a VOR/DME beacon operated by the airport on the island of Guam. The plane and the VOR/DME system were both working perfectly. The captain was a highly experienced pilot with 8,900 hours of flying time, having also flown for the Republic of Korea Air Force. He had landed at Guam at least eight times, as recently as the previous month. The other members of the flight crew were also experienced and well trained. It was raining on approach, but not dangerously so. This was a fairly routine landing conducted by highly qualified professionals. And yet this captain and crew used the VOR/DME beacon to crash KAL 801 into the side of a hill (Gladwell 2008, 178).

This flight crew had no deficiency of up-to-date technical training. They were extremely skilled at all aspects of the navigational system and with the wide variety of circumstances in which it was used. But their crash was part of a pattern for Korean Airlines, so much so that in 1999, South Korea's president declared KAL's accident rate to be a national issue, and switched his presidential flight program to the country's other airline. According to Malcolm Gladwell's account of this story, what ultimately downed KAL 801 was deference culture in the cockpit. He traces this to hierarchies built into Korean language and cultural practice. Neither the copilot nor the flight engineer felt that they could confront, correct, or even speak directly to the captain, their superior officer. The flight engineer likely realized their course was off and should have said, "Captain, you remember that the VOR/DME beacon isn't on the runway here. And it's really too cloudy to make a visual landing." Instead, he said, "Captain, the weather radar has helped us a lot" (Gladwell 2008, 216).

When KAL finally confronted the pattern of crashes, it brought in someone who focused on changing the company's culture, particularly the modes of deference that had withstood normal crew resource management techniques. He changed the flying language from Korean to English, and one doesn't have to buy the generalizations of Gladwell or some crash investigators about Korea as a rigidly hierarchical culture to see that language change would interrupt entrenched patterns and enable things to be said that one would be reluctant to say in one's mother tongue. Many have credited a range of cultural changes with ending KAL's run of accidents: as of this writing, KAL has been crash-free since 1999.

We wish to use this story to highlight a different phenomenon. Numerical information has meaning through the institutional and cultural systems in which it is created and used. The solution to KAL's crash problem was not to create proper respect and facility for numerical information, which was already extremely high with its pilots, nor to identify individual acts of interpretive error, nor even to train officers to speak openly to their captain. Rather, the solution was to enable the whole crew's active engagement in interpreting details and anomalies in quantitative data in the qualitative context of their minute-by-minute experience of the flight.

This practice is surprisingly rare, and its rarity is one of our motives for writing this book. Crises like KAL's pattern of crashes can cause people or organizations to correct flawed relations to quantitative data, such as allowing them to overcome the false sense that numbers carry their own meanings and can be handled passively. But in general, our societies have not taken this step. Data of various kinds permeate our private, professional, and political lives. We need to make a range of personal choices about our health, family relationships, education, and consumer practices. We need to choose between political candidates, join or avoid advocacy groups and social movements, and estimate the benefits of divergent social policies. We expect ourselves to have good reasons for making these choices, and we expect the same for others. In all of these spheres we regularly treat quantitative data as decisive. We underinvest in modes of qualitative interpretation, though these are often difficult and complex. We do not design institutions to put qualitative understanding on the same plane as the quantitative. We do not create nonbinary attitudes that can bring quantitative and qualitative knowledge together. And we do not treat quantitative information as always embedded in cultural systems, where the meanings of the numerical are finally decided.

These omissions are doubly dangerous in our allegedly post-truth era, one permeated not only by internet-enabled "deep fakes" and psychological manipulation but also by a supposedly general indifference to facts, or a decline of reason as eclipsed by affect (Davies 2018).

This volume addresses the role of *numerical* information as an anchor of factuality. In a stereotypical received model, qualitative arguments, cast in language, are composites of fact and opinion, while quantitative data are precise, value-free, and objective. In this popular framework, scientific knowledge emerged centuries ago from the welter of discourse and commentary that formed natural history through the continuing process of mathematization of the relationships among data elements. One result of this view is our venerable "two cultures" model (Snow [1959] 2001), now evolved into a split between STEM disciplines (science, technology, engineering, and mathematics) and all non-STEM, the "soft" human sciences, whose conclusions and very status as knowledge are always contestable.

This binary model is, of course, incorrect, as all domains of knowledge are a complex mix of the qualitative and the quantitative (six million is a deceptively precise number from the discipline of history that every European knows, if they know nothing else about twentieth-century history). And yet the binary model remains a cultural common sense: While numbers can of course be falsified and manipulated, they rest on rigorous methodologies that bring a precision that qualitative reasoning allegedly lacks. Numbers are the foundation of scientific knowledge, while language permeates the far less trustworthy worlds of politics and culture. This dualistic stereotype affects every domain of social life. In higher education, for example, undergraduates have been told to leave the subjective and supposedly impractical arts and humanities fields for the objective and efficacious STEM disciplines, whose only common feature is that they are quantitative.

The two cultures quant-qual stereotype also has social and political consequences. Valid personal decisions and policy arguments are obligated to start with data like these, and to remain grounded in them:

- The average resident of a member country of the Organisation for Economic Co-operation and Development (OECD) has a net adjusted disposable income of just under US$31,000, lives in a household with an average net wealth of just over US$330,000, and (if aged between fifteen and sixty-four) has a 67 percent chance of having a job (OECD 2017).

- UK gross domestic product (GDP) increased by 0.4 percent between the first two quarters of 2018 (Office for National Statistics 2018).
- The richest 10 percent earn up to 40 percent of total global income. The poorest 10 percent earn only between 2 and 7 percent of total global income (United Nations Development Programme n.d.).
- The global economy loses about US$1 trillion per year in productivity due to depression and anxiety (World Health Organization n.d.).
- Juveniles incarcerated for an offense are thirteen percentage points less likely to complete high school (Aizer and Doyle 2013).
- A couple weeks ago, my husband and I climbed a 500-foot volcano. His FitBit registered 51 floors; mine said 39 (JillD55 2018).
- According to Google 2018 Scholar Metrics, *Journal of Communication* with h5-index of 49 is the top publication in Humanities, Literature, and Arts (Google Scholar 2018).
- As of 2015, global temperatures had risen about 1 degree Celsius above preindustrial levels, the warmest in more than 11,000 years (Climate Analytics n.d.).

Such data are ubiquitous. Our self-understanding is increasingly filtered through them. Our work achievements are measured against numerical indicators. The value of our university degree depends on the university's ranking. Political debate revolves around numbers and targets. During the recent, and at the time of writing ongoing, COVID-19 pandemic, this reliance on numbers has only intensified, with daily reports of case numbers or debates over the "R" number dominating public debate over policy options. How do we make sense of this data, at least in the North Atlantic world?

In fact, society's relationship to numerical data is changing, and these changes form the subject of this book.

As we've noted, the most common mode of sensemaking has been to pick the quant side of the two-culture duality, and take numerical data more or less at face value. There's the weather: today it is 16 degrees Celsius in our town, and not hotter or colder. We can make similar numerical claims about the atmosphere: "Globally averaged concentrations of carbon dioxide (CO_2) reached 405.5 parts per million (ppm) in 2017, up from 403.3 ppm in 2016 and 400.1 ppm in 2015" (Nullis 2018). Numbers dominate various kinds of assessment: Paolo got a 1070 on his Scholastic Assessment Test (SAT). The University of California at Santa Barbara is ranked fifth among US public research universities, while the University of Cambridge is ranked first in the United Kingdom. The measurement of intellectual achievement and of institutional performance are, in this view, similar enough to the measurement of air temperature or CO_2 levels to allow us to take those scores as forms of fair objectivity—or at least as fairer and more objective than the alternative qualitative measures (like evaluating an applicant's biographical statement alongside their SAT score). The faith in the transformative power of data to resolve long-standing theoretical disagreements is similarly common across social and natural sciences (see for instance the so-called empirical turn in economics).

A second response to the ubiquity of the numerical dominates scholarly literature on the quantitative. That is to historicize and thus denaturalize the numerical's intellectual authority. The nature and social power of numbers have long been on the radar of historians of science and scholars of public administration (Desrosières 1998; Espeland and Stevens 1998; Hacking 1995; Nirenberg and Nirenberg 2021; Poovey 1998; Porter 1995; Power 1997; Shore and Wright 2015). In recent decades such scholars have told powerful stories about how numbers and quantitative methods were constructed institutionally and intellectually before they could assume center stage in the culture of modern science and governance. Scholars have explored the rise of statistics, cost-benefit analysis (CBA), probability, bookkeeping, audit, metrics, and risk management, and have shown how each promised objectivity, translatability, and accountability, and yet, at the same time, disavowed their social and institutional roots, and also excluded and erased the experiences, contexts, and complexities that did not fit into the specific numerical regimes of modernity. Although few of these authors address the two-cultures framework explicitly, we could say that they show that the numerical's alpha culture emerges from and depends on philosophy and history's beta culture, rather than transcending and correcting the latter.

We shall call this line of inquiry the Original Critique—to avoid repeated citations and lengthy constructions. Although there are differences between the contributions we lump under this umbrella term, it is fair to say that the Original Critique tells a story of the rise of numerical thinking that does not equate it with the advancement of scientific objectivity. Instead, the Original Critique tracks the numerical's infiltration into processes of governance and science as involving a loss and displacement of situated, informal, qualitative knowledges. It notes the ways in which the numerical undermined the conceptual means by which it could be criticized. Its general conclusion is that numbers presuppose categories which themselves encode worldviews and histories, and its imperative is that we *interpret* numbers, in context, with an awareness of history, exclusions, intentions, and goals.

The Original Critique is important and inspiring. At the same time, it implies that the numerical always wins out over qualitative knowledge. Although this strain of thought stresses quantification as a sociocultural practice, it also suggests that once a process of quantification has started, it successfully suppresses its origins and swallows up the conceptual and institutional means to articulate alternatives. Clearly this does happen. For example, the rise of CBA, based on an account that grounds degrees of welfare in "willingness to pay," renders invisible ways of valuing which cannot be reduced to monetary indicators: the only way to prove the value of these approaches is in quantitative terms. But is this ascent of the numerical generally complete?

In contrast to what the Original Critique often implies, quantification is not a one-way process, and on many topics and occasions it can be reversed. Some modes of quantitative authority have been wholly discredited: mystical "numerology" no longer guides military decisions, nor is craniometry a basis for education policy. These examples might seem frivolous, but that is, itself, telling: there is nothing less powerful than a discredited number. The history of discredited numbers casts the validity of quantification into doubt, for the public as least as much as for experts.

Long-term scholarly work to refute specific uses of quantification like scientific racism—an important strand of the Original Critique—has always been accompanied by political, religious, or other forms of *non*expert skepticism about the implications of quantified research. These can come from left, right, and center. One major right-wing mode has been the rejection of climate science, leading to the dismissal of quantitative evidence of anthropogenic global warming. When millions of Texans lost power during a severe winter storm in February 2021, Governor Greg Abbott told a national television audience that offline renewable energy had "thrust Texas into a situation where it was lacking power in a statewide basis"—ignoring the data that two-thirds of the lost power generation was from oil and coal (Mena 2021). On the left, a similar skepticism may take the form of exposing the alliance between the experts who produce numbers and the dominant ideologies of neoliberalism and capitalism.

More recently, whether on the left or right, the dizzying speed at which experts have changed their claims about COVID-19 has created new sources for public skepticism. To an extent, one might worry that such skepticism is unfounded, because it is difficult for even the best experts to grasp a rapidly changing situation. Nonetheless, there is a clear tension between a public image and rhetoric of science as dealing with numerical certainties and the reality of shifting views and heated debate about what, how, and why to count.

We have entered a period of a distinctive *fragility* of the numerical. After the 2020 US presidential election, Donald Trump continued to make "unfounded claims that he lost re-election only because of mass voter fraud, even though such claims have been rejected by judges, Republican state officials and Trump's own administration"—and 82 percent of Trump voters agreed (Associated Press 2021; Salvanto et al. 2020). This fragility appears in public skepticism about the uses to which quantitative objectivity is put. In a recent UK study, 60–70 percent of respondents thought that "official figures were generally accurate," while about 75 percent thought officials and the media did not represent these figures honestly (Simpson et al. 2015, 26–27). This skepticism about the quantitative modalities of expertise has been part of Anglophone intellectual culture for generations, and yet there have been few times when it has had the political and cultural salience it has today. It can produce a polarized standoff between two untenable positions—data denial and numerical absolutism—as when Trump was confronted by Georgia's Republican secretary of state, Brad Raffensperger, who said, "Working as an engineer throughout my life, I live by the motto that numbers don't lie."[1]

Quantitative knowledge now exists on a confusing terrain. We continue to work with the two-cultures hierarchy of knowledge, reflected in the superior authority of numerical information. Metrics and indicators continue to spread: they monitor performance, sort and rank individuals and institutions, influence policy, and shape cultural expectations. At the same time, their authority is subject to a more diverse set of persistent challenges than at any time in recent history. The challenges are politically diverse.

Making matters still more complicated, public skepticism does not align with the Original Critique of decontextualized and dehistoricized quantification. That critique's main conclusion—*interpret* numbers, in context, with an awareness of history, exclusions, intentions, and goals—seems to have been overrun by a politicized skepticism that says ignore numbers whenever you don't like their conclusion and make up alternative ones if necessary. In response, many officials proclaim, "I believe in science," countering skepticism in Raffensperger style by implying that numbers have an objective face value (Biden 2020). This position also ignores the Original Critique, and blatantly overstates the case, since the inevitable anomalies and ambiguities of science were out in the open with COVID-19 and generated regular changes in scientific policy advice.

Could numbers that govern, command trust, and serve as the basis of common policy have always been as unstable as they now seem? Were we wrong to worry so much about the "tyranny of metrics"? Can we reconnect the Original Critique to public skepticism, or do we need a new scholarly perspective to replace the Original Critique?

These questions receive additional urgency from the arrival of "big data," which may represent "an entirely different type of knowledge, accompanied by a new mode of expertise." The sociologist William Davies has argued that emerging forms of quantitative analysis are often nonpublic and lack a "fixed scale of analysis" and "settled categories":

> We live in an age in which our feelings, identities and affiliations can be tracked and analyzed with unprecedented speed and sensitivity—but there is nothing that anchors this new capacity in the public interest or public debate. There are data analysts who work for Google and Facebook, but they are not "experts" of the sort who generate statistics and who are now so widely condemned. The anonymity and secrecy of the new analysts potentially makes them far more politically powerful than any social scientist. What is most politically significant about this shift from a logic of statistics to one of data is how comfortably it sits with the rise of populism. Populist leaders can heap scorn upon traditional experts, such as economists and pollsters, while trusting in a different form of numerical analysis altogether. Such politicians rely on a new, less visible elite, who seek out patterns from vast data banks, but rarely make any public pronouncements, let alone publish any evidence. These data analysts are often physicists or mathematicians, whose skills are not developed for the study of society at all. (Davies 2017)

As these techniques infiltrate public services such as policing, education, and health care, they introduce new forms of authority over inmates, teachers, and patients, potentially changing the nature of our encounter with the state. Popular exposés now tend to emphasize the powerlessness and the arbitrariness brought by the reliance on algorithmic solutions rather than on the judgment of trained professionals, even with all their biases (O'Neil 2016). Structural forms of discrimination, such as race-based credit discrimination in the United States, may be baked into algorithms, whose status as proprietary business secrets blocks correction (Noble 2018). Do such trends mark the resurgence of what Elizabeth

Chatterjee (chapter 1 of this volume) terms the "quantocracy"? Or have we entered new conceptual territory, in which the always-advancing practical power of quantification is coupled with a new ease in denying its epistemic validity?

This is the point at which this volume's research enters the fray. We accept the value of the Original Critique but have conducted a new round of research to understand how numbers in the present political and intellectual moment might and should work in our social and political lives. From 2015 to 2019, three teams, from the universities of Chicago, Cambridge, and California, Santa Barbara, undertook to study the history and the present of quantification in three areas: climate, health, and higher education. The Chicago project analyzed the role of numerical estimates and targets in the explanation of and planning for climate change. The Cambridge team focused on numbers that are said to represent health, and the effectiveness of medical interventions and well-being more broadly. The team in Santa Barbara examined the quantification of outcomes in teaching and research, as well as in discussions of the effects of university teaching. As we grappled with the complexities of our chosen case studies, we came, in spite of our own internal differences, to recognize two key starting points for our research.

The first is that it is easy but wrong to think of quantification and the growth of metrics as a single homogenous thing. "The numerical" as an umbrella term can represent vastly different practices. A complex table of indicators by which to judge progress—as in the United Nations Human Development Index (HDI) or the Sustainable Development Goals, or the UK Office of National Statistics' Measuring National Well-Being program, discussed in chapter 8—is numerical but it is conceptually at odds with traditional economic CBA, which is also numerical. Proponents of the HDI are as "quantitative" as proponents of CBA, and were looking for a simple and manageable number to replace GDP (Morgan and Bach 2018). However, the HDI rests on a decidedly Aristotelian account of well-being in terms of beings and doings, rather than the utilitarian viewpoint that informs CBA. The latter seeks to commensurate all value in terms either of pleasure or strength of preferences, an anathema to the capabilities approach that informs HDI. For certain purposes it may be helpful to lump these cases together, but problems arise when we apply the same critique to them: they in fact deserve different treatments. On the face of it, different statistics, say of well-being in chapter 8 or of the value of education in chapter 10, may look similarly quantitative but are in fact based on dramatically different and often equally shaky conceptual foundations. When we attempt to uncover differences between numbers we discover a complex interplay of reasons to choose a given numerical technique. This should push us to give a more precise definition of "the quantitative" and a more subtle critique of it.

The second is that while the Original Critique built on Foucauldian insights to articulate the relationships between power and quantification, we are all impressed by a countercurrent of research which stresses how numbers can be used to challenge vested interests, in part by making the invisible visible (Bruno et al. 2014). This tradition, mainly based in France, is exemplified by Thomas Piketty's memorable claim that "refusing to deal with numbers rarely serves the interests of the least well-off" (Piketty 2014, 402). Perhaps the most potent and powerful form of such *statactivism* (i.e., activism using statistics) is found in the area of climate science, where quantitative tools have allowed us to grasp the damage done to the environment by industrial development. The striking statistic that the richest 1 percent holds more US wealth than all of the middle class combined was not merely rhetorical but drawn from quantitative research and motivated the Occupy Wall Street movement. While numbers can hide the workings of power, they can also be used to challenge it, prompting political movements and opening up new forms of political action. Furthermore, there are many areas of political decision-making where we have good reasons to replace "expert judgment"—which can, all too often, serve as a cloak for bias or self-interest—with quantified measures and targets; nothing is gained from holding up a romantic ideal of professional judgment as epistemically or ethically unproblematic. Precisely because numbers are *not* above or beyond politics, they can be reconstructed and redeployed for different ends. The problem is not quantification per se but the uses to which different actors put quantification. We can use numbers to hold the powerful to account, but only when those numbers are themselves democratically accountable.

Easier said than done. How do we implement this accountability?

The Original Critique has recently been addressing this question. Since 2010, what we might call a second wave of work in this field—by authors such as Cathy O'Neil, Frank Pasquale, Sally Engle Merry, Wendy Espeland, Michael Sauder, and David Nirenberg and Ricardo L. Nirenberg—has taken important steps toward recognizing how we might create epistemically and ethically responsible indicators. In this context, our volume makes several distinct contributions to the study of the numerical as a simultaneously intellectual and social issue.

First, we define quantification more narrowly and perhaps precisely than many of our colleagues, as follows:

Quantification is the deployment of numerical representations where these did not exist before to describe reality or to affect change.

Our definition emphasizes the descriptive and the active roles of numbers. It shows quantification to be a continuous project, which must always be understood relative to its goal, context, and to history. And it does not restrict the domain of quantification in the manner of other definitions.[2]

Second, we start from the contradictory or at least paradoxical terrain noted above, in which numerical data and indicators are both more influential and more fragile than before—both more and less authoritative, harder to contest and more contested. Sometimes they persist stubbornly (as in our commitment to the idea that we have exactly five senses), but at other times we discard them easily and without regret. This seems different from nonnumerical concepts which define our worldviews in deeper ways and take a bigger shift to challenge.

Third, in keeping with our understanding of the plurality of types of quantification, we offer case studies within our three terrains of health care, climate science, and higher education. We think that we should be cautious in moving too quickly from accounts of the urge to quantify as a general historical phenomenon to accounts (or criticisms) of particular cases. After all, one key feature of numerical representations—well explored in this volume—is that they often travel beyond the original contexts in which they were produced or intended for use to enter new domains. For example, Stephen John (chapter 6) traces the journey of the "five-a-day" number from a World Health Organization scientific report to health policy on a national scale. When numbers make these journeys, they lose some of their original features and gain new ones. The macro trends may tell us something about which *kinds* of journeys numbers can take, but they don't necessarily tell us very much about precisely where specific numbers will end up, or what work they might do, or, indeed, that such journeys must be bad or problematic.

Fourth, we attempt to be clearer about the historical and social constructedness of the numerical. It is now uncontroversial that what, how, and why we count is always shaped by our values and interests at least as much as the nature of the thing we are counting. But this has to be a starting point rather than an object of study to be demonstrated yet again.

Fifth, we use this manifest constructivism to call for the (re)use of the numerical rather than its repudiation. In working through our case studies, we became convinced that study in this area required a change of emphasis, in large part because attacks on clearly misplaced forms of quantification—such as citation metrics and satisfaction ratings in higher education—are more powerful when they are combined with an appreciation of what numbers and metrics can do well. More generally, we insist that humanities scholars are more likely to make their voice and criticisms heard when it is clear that their discussions are grounded on a detailed understanding of the limits—rather than categorical rejection or ignorance—of quantification.

The sheer heterogeneity of forms of quantification—and their purposes, effects, and uses—should make us skeptical of overarching theories of the numerical. This is why this volume studies quantification on a case-by-case basis.[3] At the same time, we intend to contribute to the understanding of quantification as a central intellectual, scientific, and social process, to qualitative methods for studying it, and to putting numbers to qualitative use.

How can we move from our particular cases to a more general assessment? To address this question, we will start with a blunt juxtaposition of the critiques of quantification with its defenses. We sourced table I.1 using both existing literature and findings from the chapters in this volume. In the table, "quantification" covers an intentionally wide range of phenomena—quantitative methods in scholarship, measurement in science, metrics in policy and services, appeal to numerical outcomes in politics.

We present this table with a goal of making explicit the need to move beyond the Original Critique's tendency to stress the left-hand side at the expense of the right. We insist on recovering both (to restate the fifth feature of our analysis). However, we do not mean for this table to be a method for weighing pros and cons in the tradition of Benjamin Franklin's "moral calculus." Indeed, any such attempt would be subject to many of the critiques of quantification in the left-hand column! We are also aware that many of the arguments "for" or "against" using numerical measures are themselves based on contestable moral or political theories, such as Rawlsian models of public deliberation. We intend this table to be used critically and opportunistically. For example, those who are interested in metrics of well-being would be wise to recognize both the poverty of any particular questionnaire or indicator that tries to capture the goodness of life and the fact that, absent such metrics, public debate will be conducted in terms of GDP. Those who worry about the arbitrariness of any particular target for schooling or health care should also acknowledge the role of these targets in holding officials accountable to electorates in modern societies where face-to-face justifications are impossible.

The chapters in this volume grapple with many examples of these complexities. The common vocabulary pulled together in the table enables us

TABLE I.1. Criticisms and defenses of quantification

Criticisms of quantification	Defenses of quantification
• Distorts through its failure to represent nonpecuniary value of goods and services, thus obscuring multidimensionality and heterogeneity of values.	• Represents phenomena with precision and parsimony where these are possible.
• Simplifies, accompanied by loss of interpretive complexity, local context, and qualitative experience.	• Establishes standards of causal inference for science as well as for judging effectiveness of policies, services, and medical treatments.
• Conceals value judgments behind the authority of the numerical.	• Enables public reason by providing common language to hold officials in the public sphere accountable.
• Creates perverse incentives to value numerical outcomes for their own sake.	• Provides an essential basis for regulation and expertise.
• Undermines the autonomy of, and trust in, professionals.	• Allows comparability of outcomes, coordination, and communication, both in sciences and in governance.
• Creates measurement burden and increases international inequalities when measurement and indicator production are expensive.	• Reforms entrenched professions and traditional metrics such as GDP.
• Silences and causes loss of political agency when numbers are used to bypass vernacular, standpoint, or subaltern knowledges.	• Uses hidden numbers to represent injustices: statactivism.
	• Enables archival work, retrieving "the great unread" with large scale data methodologies.

to get past simplistic distinctions between good and bad numbers, and past unhelpful Manichean splits between cold technocratic rationality and warm qualitative narratives.[4] It also allows us to recognize that epistemically problematic numbers can do useful social work, while epistemically legitimate numbers can fail to do so. Even if calculations of the "reproduction number" for COVID-19 are shot through with uncertainty, the number can serve a useful function of focusing attention on the possibility of exponential growth; precise measures of the number of COVID-related deaths might be epistemically impeccable, but far less powerful in motivating change than images of crowded hospitals. Sometimes a number needs to be converted into a more visceral one, as when the pandemic deaths are presented in terms of how many 9/11s or Vietnam Wars they correspond to. Our thought—explored throughout this volume—is that when we assess the limits of quantification, we need to be alert to the epistemic *and* the practical dimensions at the same time, recognizing that how they do—and ought to—relate depends on context and varies from case to case.

We come to a sixth feature of this volume, which is that we are working toward a midlevel or middle-range theory of the numerical (Alexandrova 2018). We propose neither a general theory nor a collection of discrete cases. We work with the fact that these different classes of uses, epistemic and practical critiques of the numeral, can be blurry for a variety of reasons. For example, when epistemically useful indicators are used as targets, we can create perverse incentives which undermine the effectiveness of these indicators. This is the story of attempts to assess learning outcomes told by Heather Steffen in chapter 3, part of the move of universities to redirect resources to achieve higher rankings. What started as a worthy goal of representation now warrants a critique in terms of practice. More fundamentally, constructing numerical measures may require us to make contestable value judgments, which then, to legitimate the measure, are hidden from view. For example, measures of inflation are based on the price of a standard basket of goods. When economists decide which goods to place in this basket, their choices reflect (perhaps implicit) value judgments about what is required for a decent or normal existence (Reiss 2016). In examining these assumptions, we provide both epistemic and ethical reasons to worry about inflation measures. Of course, there are often good reasons to distinguish epistemic and practical concerns within some context, and it may be that, ultimately, there is a deep division between theoretical and practical reason. And yet, at the midlevel we seek to occupy, these concerns interrelate in ways that problematize any simple claims to priority.

Throughout this introduction we have stressed the heterogeneity of numbers and the associated variety of forms—and critiques—of quantification. The reader might, then, be left suspecting that in trying to establish the "limits of the numerical," we have instead shown the limits of our own analytical categories: that there is no single thing, "the numerical," to analyze. Yet this would be a mistake. Clearly, representations of how the world is or ought to be framed in numerical terms differ from other kinds of representations, be they linguistic or pictorial (or graphical or musical or . . .). Most notably, numerical representations are tractable to certain types of formal analysis and, as such, often carry the sheen of objectivity associated with the formal sciences. Furthermore, they allow for commensuration of very different sorts of things, thus both allowing new forms of reasoning and eliding older forms. Finally, perhaps as a function of these other features, they can be transported across linguistic or cultural borders, allowing them a certain kind of authority in some debates and often displacing other forms of judgment and understanding. We do not deny these features so much as suggest that there is no one-size-fits-all story of how quantification happens, what its epistemic strengths and weaknesses are, and what its effects will be. Moreover, the reason why any one-size-fits-all story is too general for thinking through particular cases is that, as Trenholme Junghans points out in chapter 4, quantification and qualification are not really opposites but rather are two versions of abstraction. Both attempt to reduce reality under a label to make it more manageable, and hence many of the criticisms in our table above can be directed to language generally. An interesting corollary of the idea that all description, numerical or not, is selective is that the Original Critique is just a variety of what we might call the Primordial Critique—that all representation distorts. But for our purposes we are more eager to emphasize that, sometimes, hiding complex value judgments, displacing expert judgment, and misrepresenting complex reality are precisely what is needed!

Our common theme is methodological rather than substantive: we do not seek to quantify the effects of quantification or, on the other hand, to invert the traditional hierarchy and put qualitative approaches on top. Rather, our approach suggests that it is only by stepping outside quantitative frameworks that we can get a fuller appreciation of what those frameworks do, how they do it, and whether they do it badly or well. We also show the many ways that qualitative and quantitative approaches can productively interact—how the limits of the numerical can be at least partially overcome through equitable partnerships with historical, institutional, and philosophical analysis. Sometimes this analysis involves pointing to ways in which we might quantify better; sometimes it suggests that it would be better not to quantify at all; sometimes it has no normative implications, but rather deepens our understanding. Often, numerical indicators are associated with notions of rigor and clarity; we hope to show that qualitative approaches can provide a no less rigorous way of assessing the limits of the numerical while pointing toward better methods.

As a crude approximation, we might say that, at the start of our project, we broadly thought that quantification was a force for good in climate science, a force for bad in higher education, and a mixed bag in healthcare contexts. As our project developed, however, we soon found complexities in our narrative, uncovering, for example, epistemic problems with some core numbers in climate science, weaknesses in narrative alternatives to learning assessment, and unexpected benefits to doubtful metrics in health policy. Furthermore, through three years of discussion among members of the three groups, we came to understand the commonalities and differences among our respective fields, as well as the ways in which these fields have been differently affected by larger political changes like the rise of populism.

We have decided to integrate our three domains into dimensions of different topic areas, which form a story line for our readers. Such readers, we imagine, are skeptical about common uses of the numerical to address social, cultural, and political issues. They may doubt the validity of deciding whether to fund a drug treatment by assessing its cost per quality-adjusted life year (QALY) or the benefits of picking universities in accordance with their national rankings. They may be grateful for quantitative data about CO_2 levels that help explain growing climate instability while doubting their accuracy or their impact on public opinion or both. They may have abolitionist instincts in relation to using numbers to make either personal or policy decisions about health, education, and environmental survival. If they do accept numbers as basic empirical data, they may insist that data be interpreted contextually and be clearly subordinated to political and philosophical judgment. In short, these readers would pick up this book in order to see the numerical put in its place and qualitative, narrative processes be offered as substitutes for today's quantification order.

Our story line has a few twists for these readers. We begin with the supposed populist revolt against facts, truth, experts, and perhaps reality itself. We do not at all downplay the rise of authoritarian leaders and political cultures that act as though the truth is whatever they say it is, or the seriousness of their purging institutions of people and procedures that are independent of their will. But Elizabeth Chatterjee (chapter 1) and Christopher Newfield (chapter 2) argue that experts were themselves to blame for the discrediting of "objective" quantitative expertise, precisely by using decontextualized numbers and ignoring the social costs, only to see new authorless numbers come back with a vengeance in the populist revolt. Experts in the United States and the United Kingdom are paying a high price for separating quantitative claims from their motives and impacts; a similar mechanism may well explain developments in other countries. Whether it is possible to place numbers in their context, acknowledging gaps and uncertainties, and retain public trust is, perhaps, the biggest challenge we face.

The most reasonable response to the problems created by splitting numerical procedures from context is to reconnect them. Our next three chapters examine efforts to use narrative to correct the numerical—to use the strengths of the humanities disciplines to make a better use of the quantitative. As Heather Steffen (chapter 3) shows, this kind of return to the qualitative can in fact confirm or rationalize a preexisting quantitative claim and make it seem more user-friendly and acceptable without improving the outcomes. In her story, the auditor of teaching and learning takes charge of the environment and imposes a kind of authoritarianism of evaluation. Once the auditor has created a world, all problems can be recast as having solutions the auditor possesses, with the quantitative and the qualitative deployed for the same purpose. The return to narratives of individual can also reflect corporate manipulation and the lowering of standards of evidence. In Trenholme Junghans's discussion (chapter 4), qualitative narrative has its own aura of authenticity which can be as deceptive as the aura of objectivity that numbers exude. In Laura Mandell's study (chapter 5), we see that the narrative *can* change and improve numerical procedures, but primarily (and perhaps only) when it engages in a more complicated and risky process of explicitly introducing the investigator's subjectivity into the inquiry rather than trying to purge it. This exposes qualitative practice to the criticisms that empowered quantitative procedures in the first place. But subjectivity is at the center of what qualitative analysis adds to the numerical, and must not be avoided.

But given the risk, we may feel called to attempt yet another strategy, which is to accept that numbers are going to decide the interpretative outcomes if they are present at all, and thus concentrate on fixing the numbers themselves. Of course data quality is fundamentally important, so investigators are always obligated to have the best possible data and analytical procedures, to open these procedures to inspection and critique, and to strive constantly to improve them. But good numbers do not guarantee good conclusions; at the same time, good conclusions do not always require good numbers (see chapter 6, by Stephen John, and chapter 7, by Gabriele Badano).

Our story leads to a dual obligation. First, we need to engage with numerical processes rather than taking their outputs as given. This will immerse members of qualitative disciplines in expanded work on numbers, and require expanded quantitative skills. Second, we will also need to accept the power of interpretive frames over the numbers. The major world issues that quantitative methods address can be addressed correctly only if both data and interpretive frames are accurate and sophisticated, and if the investigator has equal sophistication about the details of their relationship. In chapter 9, Greg Lusk illustrates the moral perspectivalism of quantifying the effects of extreme weather: the same number might have radically different consequences for accountability and hence for the compensation of victims. The other two discussions, by Anna Alexandrova and Ramandeep Singh (chapter 8) and by Aashish Mehta and Christopher Newfield (chapter 10), both drive home the idea that even when the measurer acknowledges that their numbers may not be getting at "the real thing," and even when spurious indicators can be justified on practical grounds, conceptual work on what that real thing really is still needs to be done—and can be.

This volume contributes to the analysis of knowledge at a critical moment in its history. The dysfunctions of the two-cultures hierarchy are clear in both theory and practice, and yet we lack a new paradigm for their interaction. The pandemic, with its clarion calls to trust "the" science, with the humanities left to one side, suggests the powerful pull of the old paradigm, and yet we have also seen its weaknesses, as scientific uncertainty has bred skepticism. The plot outlined by our chapters leads toward this new paradigm. It will be grounded in epistemic parity between quantitative and qualitative disciplines. It will require intellectuals to understand more clearly the social and technical elements of the knowledge systems in which their work takes place, to formulate clear positions on these systems, and to debate them publicly so they can be changed to fit new needs. We hope that our work can be part of this complex international project.

NOTES

1. "Georgia Secretary of State Certifies Election for Joe Biden," *PBS NewsHour*, November 20, 2020, https://www.pbs.org/newshour/politics/georgia-secretary-of -state-to-certify-election-for-biden.

2. Compare this definition with that offered by Merry in her *The Seductions of Quantification*: "By quantification, I mean the use of numbers to describe social phenomena in countable and commensurable terms" (Merry 2016, 1). Merry's definition alludes to measurement as a key feature. This definition captures many practices: devising indicators and inference techniques, collecting and representing data, embedding them into existing or new institutions for purposes of control, visibility and governance. Such a definition served the Original Critique well in that it enabled scholars to find interesting facts and commonalities on which their further analysis rested—for example, that the drive to quantify came not from physics envy but from demands of procedural fairness and beneficence (Porter 1995), or that even when quantification fails it manages to reproduce itself (Power 1997). However, it also collapses all practices of quantification into counting, ignoring the fact that numbers are often explicitly introduced as a way of remaking the world, rather than counting what is there already. In keeping with similar concerns of these authors, we want to stress that quantification can be explicitly aimed at *changing* as well as at *describing* the world; the twist is that numbers can change the world only when they are presented as being grounded in reality.

3. In brief, we found useful four distinctions *within* the uses of numbers:

- *Ordinal versus cardinal numbers.* Ordinal numbers are meant only to rank items in an ascending or descending order, while cardinal numbers represent intervals between these items. A long history of debates in economics and psychology focus on whether quantities such as utility and affect are ordinal or cardinal (Moscati 2013). Although both are numerical, it is only cardinal numbers that license heavy-duty quantification that calculates averages, determines coefficients, and compares by how much items differ. Ordinal indicators that contain information only about the order do not warrant such operations. In the social sphere, some numbers are ordinal (life satisfaction, quality of teaching) and others are cardinal (numbers of votes, GDP). There is a constant temptation to treat ordinal numbers as cardinal ones and, although it is unjustified *even on the theorist's own terms*, numerical claims that go beyond the available evidence are rife. In Alexandrova and Singh's study (chapter 8), this process is exemplified when ordinal survey reports end up being used to calculate the precise impact of policies on the welfare of nations. This distinction is thus useful for understanding the means by which numbers get overloaded beyond their original capacities.
- *Empirically grounded versus spurious numbers.* A number is always presented as representing a given phenomenon. However, agents sometimes use numbers which are "spuriously precise," in the sense that they provide a more exact representation than is—or could ever be—warranted by evidence. Agents deploy these numbers sometimes maliciously and opportunistically but sometimes also because there are strong pressures to have a number, any kind of number. Such pressures can come from need for coordination, ease of communication, or for accountability. This coordinating role of the numerical is found in areas as diverse as dietary advice (John, chapter 6), economic debate (Chatterjee, chapter 1; Newfield, chapter 2), and the rationing of health care (Badano, chapter 7). The tension in the social and political uses of numbers arguably points to an even deeper tension between two roles of the numerical: as representing reality and as allowing coordination. In establishing fixed systems of measurement—say, the standard kilogram or the size of a bushel of corn—we are creating both new ways of representing the world and new ways of smoothing social coordination.
- *Scholarly versus managerial numbers.* Some invocations of numbers are supposed to represent the world but others are supposed to change it, whether by encouraging, incentivizing, or discipling audiences. It is typical to associate numbers in academia and scholarship with the former and numbers in policy, politics, and business with the latter. This, however, is a simplification: academic representations are always embedded in speech acts with consequences for audiences' behavior, and managerial numbers must be grounded in practices of representation—a clearly impossible target is no target at all. Nevertheless, distinguishing numbers in terms of their primary or intended function is crucial for understanding why conditions of success and adequacy for numbers can differ. Steffen (chapter 3), John (chapter 6), and Lusk (chapter 9) show how this distinction plays out in education, nutrition policy, and climate science, respectively.
- *Oppressive versus emancipatory numbers.* Cutting across the three distinctions above, the fourth distinction highlights the fact that whether the numbers are cardinal or ordinal, well-grounded or spurious, or managerial or scholarly, they can be used just as well to further progress as to stunt it. Where the Original Critique (along with W. H. Auden in his 1940 poem "The Unknown Citizen" and many oth-

ers) saw the power of statistics to homogenize, to erase individuality, to take away judgment and spontaneity, to manipulate, there is also, as statactivists show, the power of the right numbers to mobilize, to expose injustices, to give voice, and to represent. Where some would interpret this dual power of numbers as neutrality (numbers are mere tools, and it's only their use that can be evaluated), we rather saw agents' ability to harness the varied agendas of numbers in ways that cannot always be predicted based on how these numbers were used historically. These changes in the moral valency of deployment can be seen in Badano's story of the role of QALYs in fair allocation of resources (chapter 7) and in Alexandrova and Singh's rendition of the quantification of well-being for progressive means (chapter 8).

These distinctions helped us shift the standard definition of quantification.

4. A further feature of the table bears consideration. In keeping with the variety of numbers we have discussed in this introduction, the critical or justificatory considerations we list in each column have different goals. Some examine the ability of numbers to represent reality. Advocates point out their role in capturing fine-grained and precise phenomena, as well as enabling causal inference, a hallmark of the scientific method and everyday reasoning. Critics note the distorting effect of quantification on those phenomena that are vague and do not admit of precision or homogeneity. These are *epistemic* arguments because they focus on the successes or failures of numbers in their role as ways of representing or understanding the world. Other arguments attend to what numbers do in the public sphere once they are part of institutions such as audit or service provision—in particular, their ability to create new forms of authority, to set the terms of conversation, to make certain facts visible and others invisible. For want of a better term, we call these *practical* arguments. We class the first three as epistemic and the latter four as practical. The payoff is being able to see that epistemically problematic numbers can do useful work, and epistemically legitimate numbers can fail to do so.

Expert Sources of the Revolt against Experts

CHAPTER 1

Numbers without Experts

The Populist Politics of Quantification

Elizabeth Chatterjee

I think the people of this country have had enough of experts . . . saying that they
know what is best and getting it consistently wrong.

—Michael Gove, UK secretary of state for justice, June 3, 2016

"In contemporary politics," declared the economist Lorenzo Fioramonti,
later elected to office for Italy's populist Five Star Movement, "numbers
have been used to strengthen technocracy at the expense of democratic debate" (Fioramonti 2014a). This is the dominant understanding of quantification in public life. Technocratic governance by numbers promises objective, transparent, and rational policy solutions, transcending the vagaries
of individual judgment or the grubby partisan interests of elected politicians. Through this patina of impartiality, it can undercut political contestation. Such powers of depoliticization inform "the seductions of quantification" (Merry 2016, 4). The ubiquity of quantification has thus prompted
a number of apocalyptic diagnoses. Critics argue that an unelected elite
of professional experts deploying quantitative modes of knowledge—a
quantocracy—dictates policy in ever more domains, its metrics infusing a
self-policing promarket ideological consensus throughout society (Boyle
2000; Brown 2015; A. Davis 2017; Earle et al. 2017; Fioramonti 2014b). The
quantocracy's dominance hollows out formal democratic institutions, rendering them a mere façade that masks an elitist regime of "postdemocracy"
(Crouch 2004; Hay 2007; Mair 2013; Streeck 2014).

The political developments of 2016 cast doubt on such diagnoses, as a majority of the British electorate voted to leave the European Union ("Brexit")
and Donald J. Trump was elected president of the United States. Both of
these successful populist campaigns sought to discredit the parades of
statistics-wielding experts who lined up to dispute their arguments.[1] Trump
called the official unemployment rate "phony" and climate science a hoax,

for example, while pro-Brexit politicians declared that "there is only one expert that matters, and that's you, the voter."[2] Their attacks on international institutions and the global trade order directly opposed the quantocratic economic orthodoxy. Accordingly, the populist upsurge is often read as a backlash against the overweening power of elitist expertise and the technocratization of public life (Eatwell and Goodwin 2018; Kelly and McGoey 2018; Mounk 2018). Little predicted by pundits and political scientists, the rise of the Anglo-American populists raises important questions for our understandings of the sociopolitical power and prestige of quantification. If quantocracy were as powerful and totalizing as much of the above scholarship suggests, how could it prove so suddenly fragile in 2016?

This chapter argues that the public prestige of experts does not automatically rise and fall in tandem with numerical politics. As classic scholarship on quantification recognizes, the relationship between numbers, expertise, and political authority is a historically contingent one that raises complex questions about trust and democratic legitimacy (section 1.1). Well before 2016, this relationship had begun to fray. The increasing prominence of numerical technologies in public policy from the 1980s onward was in many ways an expression of expert weakness, not strength. Numerical performance indicators, measures of public opinion, and spurious numbers were examples of quantification *against* rather than *by* experts, prefiguring populist attacks on bureaucrats and other professionals (section 1.2). This gradually devalued the currency of expertise: popular critiques suggest that increasingly large numbers of citizens began to associate the political deployment of statistical expertise with opportunism, irrationality, and hypocrisy (section 1.3). While this is far from the only explanation for Brexit and Trump, it was an important contributing factor in reaching the tipping point of 2016.

Contrary to conventional belief, though, the populist attacks on quantocracy did *not* mean a total rejection of numerical politics. Drawing on evidence from political speeches and a corpus of 8,825 of Trump's tweets,[3] the second half of this chapter argues that Anglo-American populists have instead introduced their own charismatic *numbers without experts* (section 1.4). Plebiscitary numbers, such as poll results and Twitter followings, have gained dramatic visibility. Rejecting (visible) expert mediation, such metrics claim a new and more "authentic" legitimacy as direct, and often more intuitive and dramatic, representations of the popular will. They frequently foreground ordinal positioning, with its zero-sum logic of relative rank, rather than the cardinal indicators preferred by many quantocrats. Though some of the more spurious numbers expounded by Trump and Brexit's advocates have received disproportionate attention, these techniques are far from entirely new or extraordinary; a similar logic is shared with the superficially very different work done by dashboards, big-data analytics, and even the focus groups beloved of spin doctors, discussed in section 1.2. Nonetheless, such plebiscitary numbers leave the new class of politicians vulnerable to their own crises of disappointment and disaffection. Both of these trends—the breakdown of the link between quantocracy and depoliticization, and the rise of numbers purportedly free of experts—require us to update our sociologies of numbers in public life.

1.1. THE FRAGILITY OF TRUST IN NUMBERS

Demagogic populism is a symptom. Technocratic neoliberalism is the disease.

—Michael Lind, *The New Class War* (2020)

Quantification and the depoliticizing promise of technocracy have been bound together from the outset. The first generation of sociological scholarship on quantification argued that the rise of quantitative measures was initially tied to the growth of the nation-state in nineteenth-century Europe—as the very term "statistics" suggests—and its bureaucracy of expert administrators (Desrosières 1998; Hacking 1990).[4] Nonetheless, the relationship between quantification and experts was a complex one. Theodore Porter's classic *Trust in Numbers* (1995) argues that these new, rationalizing professionals embraced quantitative measures as a rule-bound alternative to the discretionary decision-making of traditional elites. The numbers that the new experts produced boasted apparent impartiality, guaranteed by rigorous procedures that lent their outcomes a patina of "mechanical objectivity" in contrast to the subjective (even if trained) judgments of the old experts (Daston and Galison 2010).

Quantification thus promised increased efficiency and fairness: optimal policy solutions could be calculated and socioeconomic wins and losses made commensurable, thereby purportedly putting the interests of the entire nation above the representation of any single interest group. More than this, it promised a more accountable and thereby genuinely democratic mode of governance. By creating a shared base of facts, it could "forge a new basis for social consensus and peace" that generated public legitimacy through its openness to scrutiny—though this also took much more authoritarian forms in the numerical governance of colonies via slave-ship logs and imperial censuses (Davies 2018, 63, 59).

Divorced from this classic literature, scholarship on rule by experts has typically rejected this early democratic potential. Denying any ideology except an "ideology of method"—a devotion to optimizing rationality and efficiency through quantitatively informed decision-making—technocracy is instead seen in explicit contradistinction to the opportunism and short-termism of electioneering politicos (Centeno 1993, 312–13). Observers swiftly pointed out that delegation to specialist quantifiers came at a high cost. A growing number of laws and rules were drawn up by face-less technocrats: independent regulators, quasi-autonomous executive agencies, and other functionaries distanced from the people. Economists in particular increasingly monopolized major policymaking bodies. The apparent transparency of their quantitative products, chief among them cost-benefit analyses, was in practice obscured by the complexity of their statistical and modeling techniques (Boyle 2000; Earle et al. 2017; Fiora-monti 2014b). Critics argued that such delegation had created a chasm—a "democratic deficit"—between the governors and the governed. Insu-lated in nominally autonomous political agencies, the quantocrats lay beyond the reach of direct democratic accountability. The handover of power to such agencies in turn induced popular political apathy, visible for example in falling electoral turnouts and party memberships (Hay 2007; Stoker 2006).

Much of the latter literature assumes that depoliticization through quantification is typically *successful*, removing issues from the domain of political debate more or less permanently. Even when the dysfunctions and failures of quantitative practices have become conspicuous, scholars argue that this has paradoxically brought only a self-reinforcing prescrip-tion for *more* quantification: more reliance on rankings, more audits, more austerity (Espeland and Sauder 2016; Power 1997; Shore and Wright 2015). The power of the quantocracy seems virtually inescapable, a one-way street to postdemocracy.

Well before the populist upsurge of 2016, though, the classic scholar-ship on quantification should have reminded us that the strategy of depo-liticization was only as strong as the public legitimacy of the quantocracy and the institutional distributions of power that undergirded it. Truly expert-run governments are rare (McDonnell and Valbruzzi 2014). The re-lationship between politicians and expert quantifiers is more often a mere alliance of convenience, in which the former are the senior partners even though technocratic agencies may manage to carve out a degree of auton-omy.[5] This alliance does not necessarily retain credibility with the wider public, however. Trust in would-be quantocrats can wane if the promise of the ideology of efficiency fails to materialize. Tempted by the appeal of alternative legitimation strategies, politicians may also renege on the alli-ance, *re*politicizing the issue of expert delegation when this accords with their own interests. The analysis in this chapter thus studies delegation to quantitative experts as a politically embedded *legitimation strategy,* one only as successful as its political support and public reputation for competent performance and the transcendence of self-interest. Just as Porter saw quantification deployed against traditional elite discretion two centuries earlier, so too the tools of governance by numbers could be turned against the new quantitative experts themselves.

Where quantocracy has been publicly visible, populist leaders are likely to attack it as part of their broader assault on the legitimacy of mainstream politics and the domination of existing elites. Nonetheless, this chapter contests the notion that populism marks a simple triumph of emotion, dramatic narrative, and appetites over numbers, as conventional interpretations of Trump as the "reality TV president" suggest (Poniewozik 2019). Populists are not entirely hostile to the politics of numbers. After all, technocracy and populism are reactions to surprisingly similar diagnoses of the opacity and venality of mainstream party politics and traditional policymaking (Bickerton and Accetti 2017). Both reject mainstream modes of political mediation and bargaining or deliberation, instead claiming to offer more authoritative and legitimate decision-making.

It should be no surprise, then, that a populist politics of numbers can emerge, laying claim to quantification's allure even while jettisoning the intermediary "expertise class" itself. In place of the tarnished authority of expert proceduralism, it seeks legitimacy much more directly by appealing to the will of the people itself. The credibility of such numbers without rigorous quantification is likely be tenuous and short-lived. Nonetheless, (faux) metrics that purport to offer unmediated and commonsensical numerical representations of the popular mind have proved to be an overlooked hallmark of contemporary Anglo-American populism.

1.2. REPOLITICIZING QUANTOCRACY: EXPERTISE AND POLITICAL EXPEDIENCY

> What governs our approach is a clear desire to place power where it should be: increasingly not with politicians, but with those best fitted in different ways to deploy it. This depoliticizing of key decision-making is a vital element in bringing power closer to the people.
>
> —Lord Falconer, UK secretary of state for constitutional affairs, 2003

Long before Brexit and Trump, the strategy of depoliticization through expert delegation was reaching exhaustion, in large part because politicians had themselves repeatedly undermined it. Its fragility lies in the reasons that politicians appealed to quantocracy in the first place: to counter right-wing attacks on state inefficiency. This tactic was always a problematic one. Delegation to experts produces the very sprawl and unaccountability of the "administrative state" that right-wing critics condemned. To counter such accusations, mainstream politicians began to deploy *quantification against experts* themselves, through regimes of numerical performance targets and appeals to indicators of public sentiment. The link between quantification and expertise was thereby fundamentally reshaped in ways that echo the contemporary populist logic of quantification: experts were demonized even as governance by numbers expanded. At the same time, the numerical targets that gained traction in policy discourse were often produced not through robust expert quantification but through expedient political selection—an opportunism that diminished public trust. In these ways, quantocracy became repoliticized in the years before 2016 and its legitimacy waned.

The size of bureaucracy exploded in the aftermath of the Great Depression, wartime mobilization, and the ensuing growth in social welfare programs. Public faith in politicians' competence and altruism faltered with the stagflation of the 1970s, however, and the powerful interpretation of these problems put forward by Ronald Reagan and Margaret Thatcher: "government is not the solution to our problem, government *is* the problem."[6] This "bad faith model of politics" cast all political actors as incompetent, selfish, and/or corrupt, and the state apparatus as prone to bloat and capture (Flinders 2012, 154–59). Trust in government began to slide, and has never recovered on either side of the Atlantic.

In response—and apparently internalizing this "bad faith" critique of their own characters—politicians began to embrace strategies of depoliticization, borrowing quantocracy's veneer of objective, nonideological, performance-based legitimacy. Ditching doctrinal purity and slashing the federal workforce far more dramatically than the antibureaucratic Reagan, Bill Clinton's "third way" rested on his self-portrayal as "a technocrat who professes strict adherence to the stern discipline of the numbers and little interest in who is to blame for current maladies" (Skowronek 1997, 446). In Britain, Tony Blair's famous governing mantra was "what works," and government ministers overtly declared their commitment to depoliticization (Burnham 2001). Among their successors, David Cameron was the self-proclaimed "new Blair," a pragmatic synthesizer under whom purportedly "evidence-based policymaking" thrived; despite his transformative campaign rhetoric, Barack Obama was famously cerebral and problem-oriented as a president. These politicians thus counterintuitively reduced the potential for collective deliberation and their own discretionary power in order to enhance their governing credibility while deflecting blame for policy failures. Their proud declarations of the intent to depoliticize belie the postdemocracy thesis that there was something secretive and nefarious about this trend: in fact, it was a strategy meant for public consumption (Hay 2007).

The strategy of delegation to experts did not succeed in silencing doubters, however. Right-wing critics of the bungling government—and its contradictory cousin, the all-powerful totalitarian state—rapidly turned their ire on the "alphabet soup agencies" within which quantocrats proliferated. In the United States, fulminations against the hypertrophic managerial elite have been a mainstay of Republican politics for decades, intensified by pugilistic media coverage. On the left, too, quantocrats were demonized as neoliberal stooges, tools for the marketization and corporatization of the state. To both the left and the right, the quantocrats appeared to form a recognizable new oligarchy—the most politically powerful incarnation of what Christopher Newfield (chapter 2 of this volume) calls "the expert class," a college-educated elite intent on policing its own boundaries through education, credentialing, and cultural capital, and equally intent on protecting an economic system that served its own interests while leaving the traditional working class out in the cold. Politicians thus tried to have it both ways. They borrowed numerical tools as a means to reground public trust in state performance. At the same time, they attacked the much-loathed expert class.

At times, quantification was turned against experts. Politicians attempted to appropriate the quantocratic veneer of legitimacy while performing their own disapproval of the unpopular expert class. In line with a philosophy of economic rationality, governments on each side of the Atlantic embraced the tenets of what became known as the "new public management," especially visible in the proliferation of performance indicators, rankings, and targets. These numerical tools were often turned against professional experts themselves: doctors, teachers, academics, and the state's own vast apparatus. The Clinton administration's Reinventing Government initiative purported to reorganize and deregulate government agencies on money-saving "postbureaucratic" lines, an effort redoubled by George W. Bush, the self-proclaimed "MBA president." In Britain, the Blair government introduced a plethora of star ratings and quantitative pledges across a variety of social policy domains, most notoriously in hospitals and public schools. Most recently, this trend has accelerated as the "new empiricism" of evidence-based policy, performance management, and big data spreads in sectors from the environment to international development (Kelly and McGoey 2018). Such strategies undermine the older link between quantification and specialist professionals: they seek to minimize expert judgment and mediation, substituting instead the apparent transparency and impartiality of numbers. Numbers could be trusted, they signaled, but experts themselves must be disciplined.

Alongside the regime of numerical targets came another, apparently contradictory set of numbers: qualitative and increasingly quantitative indicators of public opinion, through which politicians attempted to emphasize their proximity to voters and transcend dependence on expert advice. Bill Clinton and Tony Blair both granted focus groups and opinion polling a famously central role in their campaigns and early years in government: public opinion data was seen to drive Clinton's hardline attitude on welfare policy reform, for example (O'Connor 2002). The rise of big data has only accelerated this trend, culminating in Hillary Clinton's unsuccessful 2016 presidential campaign. Two years before the election, her team began focus groups to discover a "core message" for her presidential run, eventually substituting for an overarching narrative a shopping list of policies tailored to every conceivable constituency (Allen and Parnes 2017). Her campaign schedule was similarly shaped by algorithmically driven data analytics, simulations, and even randomized controlled trials.

It may have won votes in the short term, but the resulting government by public opinion indicators proved deeply unpopular, branded weak and unprincipled. Politicians appeared to act like "machine-learning algorithms," to borrow an analogy from mathematician Cathy O'Neil, switching views and policies with the winds of public sentiment. In the heroic ideal of policymaking, quantocracy promises to expertly calculate a single, optimal recommendation. The pliability of public-opinion-guided policymaking directly contradicted this promise. It is no coincidence that "U-turns" and "flip-flops" have become their own genre of political fiasco in recent years. More than this, the democratic character of such polling-driven government was questionable. Like the plebiscitary numbers favored by populists (see below), such technologies of public preference revelation draw on their apparent epistemological proximity to the popular will as the source of their legitimacy, in contrast to the detached objectivity of quantocrats. Yet in contrast with the populist preference for majoritarian numbers, such research was animated by the importance of securing the centerground. Focus groups and data analytics concentrated on the median voter—caricatured as "Mondeo man" after Ford's innocuous family car—and key groups of swing voters. The core voter base could meanwhile be taken for granted, as in Hillary Clinton's catastrophic neglect of the apparently safe state of Wisconsin. In the stratified electioneering of this mainstream "market populism," an astute observer of New Labour noted, "all voters may be created equal but some have become more equal than others" (Wring 2005, 179). Not for nothing were politicians such as Blair and the Clintons self-professed technocrats, embracing governance by numbers, and simultaneously regarded as deeply unscrupulous.

Worse, for all its claims to capture the popular will, the mediation of quantocrats proved inescapable. Each of these research techniques was to be administered by professional pollsters such as the New Labour polling guru Philip Gould and his business partner, the Clinton adviser Stan Greenberg. Public opinion research was the flipside of political marketing by newly ubiquitous "spin doctors." As this suggests, it was never clear whether such governance intended to track public opinion or to mold it into a more convenient shape. The result was that politicians appeared *too* responsive to factors beyond high-minded principle or the national interest, but increasingly *un*responsive to the concerns of a swath of their traditional electorate bases (see Newfield, chapter 2) and obsessed by political marketing at the expense of policy substance. This combination of complacency and populist indulgence was a hallmark of 2016's doomed mainstream campaigns, and especially the decision to call the Brexit referendum in the first place.

Where they found the besieged quantocracy less than tractable, politicians also simply conjured up their own numbers. Quantitative expertise does not always (or even often) shape policy. Instead, expert knowledge is deployed opportunistically, especially in policy areas of high public salience. As Christina Boswell (2009) observes of immigration policymaking in the European Union (EU)—a sector that impinges directly on the "neoliberal agenda" of free trade supposedly reserved for quantocrats— politicians make decisions first and only then call on expert knowledge to lend authority to their policy preferences. Such political expediency often drives the selection of the numerical targets. In 2010, for example, the British government established an absurdly strict target to reduce net migration from the hundreds of thousands to the tens of thousands, later crystallized as a suspiciously round 100,000. This was guided less by the target's plausibility as a tool of organizational management—it was essentially impossible to meet within the framework of the EU—than the public commitment it signaled to addressing immigration (Boswell 2018). The failure to meet this target became a staple of the campaign to exit the EU, with Brexiteers arguing that impossible targets were corrosive of public trust in politicians. As the statistician-sociologist Alain Desrosières (2015)

Figure 1.1 Spurious numbers: The "Vote Leave" campaign bus. Exeter, Devon (UK),
May 11, 2016. Photograph: @camerafirm / Alamy Stock Photo.

warned, quantifiers are themselves vulnerable to unexpected feedback effects, as their own public numbers escape their control.

This expedient relationship between expertise and politics was prominently on display during the Brexit campaign, and not only in the Vote Leave campaign's notorious "Brexit bus" advertisements promising an exaggerated £350 million per week to be recovered from the EU (figure 1.1). The Remain camp also opted for the so-called Project Fear strategy, parading a string of experts to make "a naked appeal to the emotions, deploying unverifiable figures about the economic consequences of leaving the EU that were plucked from the air" (Gray 2017). The specious numbers deployed by populists in 2016—whether the Brexit bus figures or Trump's allegation that three to five million fraudulent votes were cast—were therefore hardly the extraordinary novelties that those who talk of a new "post-truth era" claimed.[7] Political numbers have long sought to exploit the credibility of rational, purportedly expert-produced statistics while abandoning the inconvenient procedural restraints that govern their production. After all, the bestselling statistics book of all time remains Darrell Huff's almost seven-decades-old *How to Lie with Statistics* (1954). Perhaps it should not surprise us, then, that a Cambridge-based research project found that 55 percent of the British population believes that the UK government "is *hiding* the truth about the number of immigrants in the country" (Davies 2018, 63). Surveys in the United States have similarly found that a majority believes the national census systematically undercounts the country's population (YouGov 2017a).

Finally, of course, politicians attacked quantocrats directly. In the United States, quantocracy became polarized on party lines, presaging Trump's outright assaults on expertise. In Britain, where the trend of depoliticization through expert delegation was more deeply entrenched and bipartisan, repoliticization was nonetheless visible. Far from silencing dissent, the ubiquitous "quango" (quasi-autonomous nongovernmental organization) sporadically came under fire in the media throughout the 1990s and 2000s for its purported inefficiency and threat to democracy (Deacon and Monk 2006). As on the other side of the Atlantic, politicians collaborated in the delegitimization of the quantocrats. Upon taking office, David Cameron's administration initiated a "bonfire of the quangos," an expensive purge-merger of over a hundred such bodies. Many of his fellow Conservative politicians leveled their own guns against the EU. Depoliticization had failed. Delegation to expert quantocrats itself became an erratically but explicitly political issue.

Well before 2016, mainstream politicians' expedient relationship with quantification—through numerical regimes that targeted experts, the quantification of public opinion, specious numbers, and the demonization of experts—had a corrosive effect on public trust across the political system. Numerical technologies of performance management internalized antistate critiques and were aimed at reducing rather than expanding the mediation of experts. By opportunistically undermining the ideal of the quantocracy while simultaneously stoking populist mobilization, the mainstream political class helped to sow the seeds of the populist anti-expert critique. In their reliance on polls and big data as windows on public opinion, mainstream politicians prefigured the preferred numbers of populists themselves. And in peddling spurious numbers and criticizing government agencies, they further devalued the currency of quantocracy.

In understanding the quantocratic project's failure, a large part of the explanation lies with mainstream politicians. The numerical idiom became just another part of the degraded rhetoric of politics, to be treated with the same public skepticism as any other political proclamation. We turn now to examine whether the British and American public truly did become skeptical in the years before 2016. We see not only evidence of publics already deeply disillusioned with the numbers peddled by experts but also that the quantocrats must shoulder a portion of the blame.

1.3. THE DISAPPOINTMENTS OF QUANTOCRACY

Americans might look back on the last 50 years and say, "What have experts
done for us lately?"

—Glenn Reynolds, "The Suicide of Expertise" (2017)

Even while much existing scholarship demonstrates that the rational
promise of quantocracy is often illusory, scholars continue to emphasize
the resilience of popular and political trust in numbers. In this literature, laypeople are generally treated as uncritical numerical consumers:
statistics are a "fetish" that the public treats with "reverential fatalism"
(Best 2012, 160); they have "faith . . . in statistical expertise" and "tend
to sanctify" the numbers produced by audits (Merry 2016, 4; Strathern
2000, 8). In reality, public trust in numbers is not static or unconditional.
Quantocracy promises efficiency and competence; popular trust in numbers depends on its performance. In the lead-up to 2016, quantocracy
instead delivered a stream of highly visible policy blunders, fiascos, and
scandals—and one catastrophic economic crisis—all while failing to deliver wage growth. Even worse, the incompetence and irrationality of systems guided by numbers was visible for all to see. The vast gap between
the myth of quantocratic efficiency and the messy and emphatically political reality also engendered democratic disaffection.

Against the myth of public sanctification, many sociological studies of
quantification show that the producers and consumers of metrics themselves recognize the flaws and ineffectiveness of the metrics with which
they work every day. Their attitude is often one of clear-eyed disenchantment, even if outright rejection seems impossible. Wendy Espeland and
Michael Sauder show that law school deans loathe university rankings and
believe they are deeply problematic but organize their behavior around
the rankings nonetheless; given this discrepancy, a "culture of cynicism"
is inevitable (Espeland and Sauder 2016, 8). The late Sally Engle Merry,
the leading anthropologist of numerical indicators, noted that such metrics continue to be used as the basis for political decisions, "even though
the users recognize that these simplified numerical forms are superficial,
often misleading, and very possibly wrong" (Merry 2011, S87). Two decades
ago, the accountancy theorist Michael Power (1997) argued that audits
are mere "rituals of verification," incessant, meaningless, and producing
organizational dysfunction. Their logical outgrowth is the ironic meta-
audit: *The Metric Tide*, an evaluation of the British university system's vast
research assessments exercises launched by the universities minister, gently critiques the destructive effects of metrics and monitoring on higher

education and research (Wilsdon et al. 2015). Among the most highly educated segment of the workforce, then, distrust of numerical governance is endemic.

Living and working with metrics is not only the privilege of law school deans and professors, however. Their flaws are perceived by a far wider public living with indicators and metrics every day. Most of today's petty number-crunchers operate not in the lucrative commanding heights of technocracy but in "bullshit jobs" that generate an endless stream of clerical outputs and reports required by rituals of measurement, assessment, target setting, and strategy development (Graeber 2018). This absurd and monotonous paperwork was literalized in *The Office*, a sitcom popular on both sides of the Atlantic. "Computer says no," similarly ran a catchphrase in the comedy series *Little Britain* (BBC, 2003–5), mocking the nonsensical unresponsiveness of petty bureaucrats blindly following algorithms. The phoniness of quantocratic regimes of management is thus widely and publicly acknowledged. Though their operations are often patently absurd, these regimes do not require true belief in order to secure their effects, as the case of Espeland and Sauder's law school deans shows: through such complicity the system perpetuates itself (see Wedeen 1998).

If resistance to workplace quantification has yet to materialize, popular critiques of political quantocracy and its inadequacies are increasingly visible. Satirical television shows like *Veep* (HBO, 2012–19) often depict politicians attempting to manipulate expert policy advice within a chaotic and laughably dysfunctional system. *The Thick of It* (BBC, 2005–12) featured a chaotic British government department trying desperately to spin several quantocratic snafus, like the premature publication of botched crime statistics and the wiping of a hard drive containing immigration records for over 170,000 people. The London Olympics mockumentary *Twenty Twelve* (BBC, 2011–12) was effectively an extended riff on the inadequacy of policy planning and the absurdity of political marketing campaigns. More darkly, *The Wire* (HBO, 2002–8) showed a police department deliberately gaming the "clearance rate," a crucial accountability metric that politicians insisted improve, by refusing to investigate tough cases or even ignoring corpses altogether (Muller 2018, 1). This is a world away from the striking intransigence of the 1980s bureaucracy in *Yes, Minister* or even the perfect quantocracy of *The West Wing* (NBC, 1999–2006), in which the US president is a Nobel Prize–winning economist. Clearly there is no simple relationship between such dramatic representations and the views of the public at large. Nonetheless, the ubiquity of the bad-faith model of politics in contemporary mass media bespeaks a crisis of faith in the wider quantocratic system.

The crisis of faith in "apolitical" quantocratic solutions to antistate critiques was a slow-burning one. Signs of disillusionment emerged early. Though both Clinton and Blair would manage reelection, both saw significant drops in voter turnout. Their administrations wracked by scandals, both men were considered distinctly untrustworthy by much of the electorate. Both left behind political parties struggling for a coherent identity in the rush to embrace the quantocratic regime of free trade and finance capitalism, embodied in the North American Free Trade Agreement, the City of London, and European integration. For all the talk of party polarization, this was a bipartisan embrace that appeared more like the work of a political cartel. For all the promises to pare down government, the aftermath of 9/11 left the state more bloated than ever. In retrospect, then, the sheen of quantocracy was tarnished even before 2008.

With the global financial crisis, the worst to strike since the Great Depression, discontent with quantocracy burst into the open. It marked a catastrophic failure of economic expertise: as Queen Elizabeth II plaintively asked economists, "Why did nobody notice it?" Solutions to the crisis, too, gave the lie to the quantocratic ideal of independent agencies emphasizing discipline, rationality, and predictability. The massive state intervention that followed—quantitative easing and the bailout of huge banks and insurers—appeared "irreducibly political," a failure to discipline the rapacious corporate oligarchy (Tooze 2018, 10). At least the US economy stuttered back to life. The Eurozone decision to adhere to the disciplinarian quantocratic orthodoxy on top of the early bailouts would prove even more destructive, condemning much of the continent to a decade of economic pain.

Critics at first lamented that this was a "status quo crisis," out of which professional economists, their agencies and prescriptions appeared to have arisen stronger than ever (Helleiner 2014). Yet, although the mainstream compact between politicians and quantitative experts staggered on, with hindsight it is obvious that public trust in quantocracy did not recover. The fragile but official economic recovery revealed the disconnect between common indicators, especially GDP, and "common sense." This was precisely the tension between the inevitable simplifications of quantification and the accurate representation of the world in all its experiential complexities that the Original Critique of quantification had highlighted (see the introduction to this volume). Pointing out this gulf between numbers and lived experience, Trump agreed that "our country does not *feel* 'great already' to the millions of wonderful people living in poverty, violence and despair" (Twitter, July 27, 2016; emphasis added). Two-thirds of his supporters distrusted federal economic data, along with one in four Democrats (YouGov 2017a). Quantification promises the precise and accurate reflection of reality, but this did not match the experience of many individuals and communities obscured beneath great numerical aggregates.

Unsurprisingly, then, a segment of the electorate began to show signs of discontent. In the United States, the Republican Party was fractured by the bailout decision. The resulting Tea Party agitations and congressional gridlock laid the groundwork for Trump. In 1958, 73 percent of Americans said they trusted government to do what was right always or most of the time. By 2011, with wages stagnant for years, the figure was only 17 percent. Britain's embrace of austerity marked the vicious high noon of a quantocratic consensus that appeared increasingly antagonistic to popular interests. The surprisingly tight Scottish independence referendum decision of 2014 showed that the backlash was already gathering steam, indicating the limited traction of fear-based economic arguments. Pro-independence voters instead cited "disaffection with Westminster politics" as by far their biggest motivator, followed by social-democratic support for the National Health Service (Ashcroft 2014). The whisker-thin populist victories of 2016 marked a tipping point in this rejection of quantocracy, not a sudden revolution.

More damaging than its dramatic failures may have been the quotidian associations that evolved between quantocratic techniques and an inefficiency bordering on farce. The illogic of systems of metrics and "economic rationality" is often highly visible in practice. Nothing encapsulates this better than the absurdist (and not entirely untrue) anecdote that perennially circulated in Britain: that the EU regulates the curvature of bananas (figure 1.2). Against its promises of competence, here technocracy is equated with pointless regulations, a stultifying obsession with measurement for the sake of it and against all common sense. The American equivalent may be the grotesque ritual of the annual tax return, an infuriatingly complex and stressful reminder of the vast cost and inefficiency of government. In this world of paperwork, quantocracy's promise to deliver rational governance could not be more clearly belied. Here the satire offered by *The Office* and *Veep* fused with critiques of the quantocratic project in its entirety.

In retrospect, then, the sense that the quantocratic status quo had survived more or less intact was premature. The 2016 results revealed electorates polarized on educational lines: those without a university education were especially likely to reject the options endorsed by binders full of economists (Goodwin and Heath 2016; Newfield, chapter 2). Surveys found that pro-Brexit voters were less likely to believe every category of

Figure 1.2 The absurdity of the quantocrats. *The Sun*, September 21, 1994.

expert, and only 14 percent trusted economists (compared to a hardly impressive 35 percent among Remainers)—behind historians, weather forecasters, sports commentators, and nutritionists (YouGov 2017b). In the United States, trust in banks similarly plunged by 22 percent over the decade after 2006, along with trust in the mainstream media historically seen as the gatekeepers publicly disseminating expert knowledge. The election results themselves helped to exacerbate this crisis of faith in the quantocratic system, as first pollsters' complacent predictions and then economists' doom-laden forecasts of the consequences were both proved wrong. Underlying both choices was what Alan Finlayson (2017) called "a kind of negative egalitarianism": not "We're all in this together" but rather "'You're as clueless as we are,' with the implication that 'Since we know what we don't know, you are the bigger fool.'"

Political scientists continue to debate whether democratic disaffection with mainstream politics has been produced more by inadequate government performance or structure (the supply side) or rising—and unmet—public expectations (the demand side) (Hay 2007; Norris 2011). Quantocracy links these dimensions: it raised expectations by overpromising rational governance, it failed to deliver on these promises, and it seemed entirely unaccountable for this failure. This analysis contends, then, that one key explanation for the populist "backlash" lies in the obvious and widening gulf between the heroic ideal of quantocracy—the regime of "what works"—and the disenchanting reality. Against the promise of governance by the numbers, the 2008 meltdown and everyday experience both revealed a regime of incompetence, political interference, and elite bias.

1.4. POPULIST QUANTIFICATION: PLEBISCITARY NUMBERS

"Give them the old Trump bullshit," he told the architect Der Scutt before a presentation of the Trump Tower design at a press conference in 1980. "Tell them it is going to be a million square feet, sixty-eight stories."

—Marie Brenner, "After the Gold Rush" (1990)

In the months after the two election results, commentators pointed to some of the headline-grabbing numbers of the Vote Leave and Trump campaigns as evidence that a new "post-truth" era had dawned. In contrast, the previous sections have demonstrated that precursors to a populist politics of quantification were visible well before 2016. The Anglo-American populists did not invent spurious numbers. Trump and the Brexiteers may have deployed obviously misleading numbers with unusual frequency and enthusiasm (former UK Independence Party leader Nigel Farage even claimed to have restarted smoking because he distrusted expert health warnings). Nonetheless, the "dark arts of numerical deception" in public life had a long history, as a series of books had pointed out (Best 2004, 2012; Blastland and Dilnot 2007; Huff 1954; Seife 2010). Numerical obsession with migrants did not begin in 2016, as Britain's net-migration targets and the old rhetoric of Enoch Powell attest (Hampshire 2018). Nor did the populists invent attacks on experts. The mainstream embrace of numerical performance management and public opinion measurement internalized assaults on the competence of expert professionals. Popular mistrust of expert quantification was already visible. The anti-expert attacks of the Vote Leave and Trump campaigns were a symptom of the waning of quantocracy, not its cause.

What, then, is new about the populist politics of numbers? Visible in both campaigns was the embrace of distinctive guiding principle of quantification. Rejecting the mainstream focus on the median or swing voter and the political centerground (section 1.2), populists have instead foregrounded *plebiscitary numbers*. They read crowd-based indicators—rally size, referendum and poll results, Twitter followers, television ratings, and even the stock market—as a signal of an amorphous general will and the source of their legitimacy. Against the median voter theorem, the logic is something more like pure majoritarianism, coupled with a heavy dose of braggadocio.

Plebiscitary numbers are particularly evident in the discourse of Donald J. Trump. Of 8,820 tweets posted between June 16, 2015, and May 28, 2017, a total of 502 explicitly refer to polls and polling.[8] This is by far the most substantial use of numbers in the Trump corpus. (It also appears to have been one of the easiest ways to earn a retweet.) Alongside polling numbers are further indicators of Trump's purported popularity. His television appearances were often claimed to bring record ratings for the program or network in question; Trump described himself as a "ratings machine" (Twitter, January 6, 2017). "Crowd(s)" and "rally" or "rallies" together recur in the same Twitter dataset 311 distinct times.[9] Such references are often seasoned with a nonspecific quantitative adjective, often a superlative—including "big," "massive," "biggest ever," and "record"—or estimates of the actual crowd size. Qualitative means of stressing crowd size were also deployed, such as the length of waiting lines, sold-out tickets, and the search for larger venues. Unsurprisingly, this type of quantitative boast also recurs in the context of social media itself, with the number of Trump's Twitter followers or the number of retweets. For the Brexiteers, the key numbers are the 52:48 percent verdict of the referendum itself. The "Remoaners" of the 48 percent are told to shut up and respect democracy.

Trump used these indicators of his popularity to compare his success with that of his opponents: "I had thousands join me in New Hampshire last night! @HillaryClinton had 68. The #SilentMajority is fed up with what is going on in America!" (Twitter, July 17, 2015).[10] This tendency is especially pronounced in Trump's treatment of critical media. He has dismissed newspapers and television channels from the *New York Times* to Fox News—before its belated and uneven conversion to his cause—with reference to their declining popularity. These metrics are rarely specific, but refer more to accusations of broadly falling ratings or sales: "Isn't it funny that I am now #1 in the money losing @HuffingtonPost (poll), and by a big margin. Dummy @ariannahuff must be thrilled!" (Twitter, July 25, 2015). In contrast, other tweets stress Trump's own wealth and efficiency. In the Trumpian mode of prosperity-gospel populism, financial success is a proxy for popularity and thus legitimacy.

This use of the numerical is a decidedly different phenomenon, however, from the reliance on expert-produced statistics. If the classic exemplar of expert numerical fact production is the accountancy technique of double-entry bookkeeping (Poovey 1998), the equivalent trope here might be the stock market. It is perhaps no accident that the stock market is one of Trump's favorite indicators. During the campaign, for example, he accused the department store chain Macy's of hypocrisy for condemning his anti-immigrant rhetoric when it was fined for racially profiling African Americans: ".@Macys stock just dropped. Interesting. So many people calling to say they are cutting up their @Macys credit card" (Twitter, July 2, 2015). Since 2016, too, the climbing stock market has been a pillar of Trump's claims to be improving the economy (and for the Brexiteers, more modestly, the British economy's failure to fall off a cliff). He regularly tweeted the stock market's progress, as on October 12, 2020: "STOCK MARKET UP ANOTHER 300 POINTS—GREATEST LEADING INDICATOR OF THEM ALL!!! DON'T RUIN IT WITH SLEEPY JOE!!!"

Accountancy requires meticulous record-keeping and a level of skill, the credibility of which is guaranteed by the application of a set of shared techniques across time. Its closest analogue is perhaps the set of accounting conventions that inform expert calculations of GDP. The stock market, in contrast, is the realm of animal spirits across multitudes. The information it provides is effectively *crowdsourced*.[11] The divergence of the stock market and the real economy during the COVID-19 pandemic—a so-called K-shaped recovery—showed just how narrow this "crowd" is. But the number was taken sufficiently seriously that early stock market woes prompted the administration to push through the largest government stimulus package in US history, the CARES Act; with the markets suitably buoyed, it remained much less interested in a follow-up that might actually help the working masses.

This logic is different to the public opinion research commissioned by, say, Clinton and Blair, designed to help gather information and tailor their messages. While Trump avidly follows polls, he is emphatic that he does not rely on pollsters: other candidates "pay these guys two hundred thousand dollars a month to tell them, 'Don't say this, don't say that'. No one tells me what to say" (quoted in Lepore 2015). Instead, crowdsourced indicators provide an alternative, outward-looking source of legitimacy. If scholarship on quantification sees the legitimacy of numbers as linked to their objectivity—guaranteed through the credentials and self-policing of disciplinary experts—populist legitimacy is a more imprecise and yet equally numerically based kind. The statistically informed expert consensus could be overruled by measures of popularity. Trump's fixation on his loss of the popular vote in the 2016 and 2020 elections, which he blamed on widespread voter fraud, and his 2016 inauguration crowd size, which prompted the coining of the phrase "alternative facts," must be seen in this context. They are an issue not just of narcissism, as was often mockingly diagnosed, but foundational to his political authority.

As these numerical examples show, their *relative* popularity is crucial in the self-definition of the Anglo-American populists. A scholarly consensus holds that populists are distinguished by their moralistic imagination of "the people" as a monolithic unity, rejecting alternative representative claims—a thesis encapsulated by their claim that "We are the 100 percent" (Müller 2016). This claim sits awkwardly with Trump's keen sense of his presidential approval ratings or the opportunistic maneuverings of elite pro-Brexit politicians like Boris Johnson. For such conveniently nonideological operators, "winning" was all. The competition and their narrow electoral successes were central in forging the movements' identities. For this reason, popularly sourced numbers are essential because they can be easily and intuitively ranked in a zero-sum logic of winners and losers (one of Trump's favorite words).

The problem with such plebiscitary numbers is that, unlike expertly produced statistics or even market research, they provide very little policy-relevant information. Tweets and crowd size privilege "the base," smaller communities of the politically active faithful, ignoring huge segments of the population. Approval ratings capture a broader range of public opinion, but they cannot guide the complex process of policy planning and implementation. They reflect only an endless popularity contest—or, perhaps, a beauty pageant—in which short-term jockeying for position is crucial. After the celebrations, then, both ensuing administrations would lapse into an unstable mode of permanent campaigning, while policymaking remained chaotic and ineffective in practice.

Behind the scenes, the mediation of experts has not ceased, but shifted toward the purportedly apolitical figure of the data scientist. Mining reams of social media data, both the Brexit and Trump campaigns used big-data analytics in order to precisely tailor messages for potential voters—and, more pointedly, to discourage opponents from bothering to vote at all. The effectiveness of this is easy to overstate; in practice campaigns often rely on surprisingly crude algorithms and datasets (Hersh 2015) and, after all, Hillary Clinton still lost despite a sophisticated big-data operation. Yet, even if their political masters and figureheads themselves remained sloppy and anarchic, a Leninist subset of quantocratic populists hoped to embrace quantification while rejecting the mainstream expert-politician relationship. Trump's son-in-law Jared Kushner, who oversaw the campaign's digital efforts in 2016, attempted to use big data to game media coverage of the administration's Middle East strategy. The éminence grise of Brexit, prime ministerial adviser Dominic Cummings, sought to turn big data into a veritable philosophy. He planned to create a Downing Street data analytics unit to exploit the government's hefty parliamentary majority, centralizing control away from the tedious and self-serving "hollow men" of the civil service. The black box of big-data analytics via proprietary algorithmic technologies offered such would-be rulers the alluring prospect of linking the leader and the voter more directly. In this vision, technocracy and populism fuse in the notion of a reengineered society without the tedious bureaucracy of the mainstream political system.

In practice, though, such visions proved as utopian as the old quantocratic promise of efficiency and irreconcilable with the broader logic of populism. A lot has happened since the first draft of this chapter—not least a pandemic and a presidential election. Much analysis still remains to be done on the relationship between those two historic events. It seems clear that COVID-19 did less to undermine support for Trump than predicted by the pollsters, those failed quantitative experts of the media. Perhaps the new data analytics helped to improve turnout amid the "base" to record Republican highs, while successfully microtargeting Cuban and Venezuelan expatriates in Miami-Dade County and the proud Tejanos of the Rio Grande Valley. Perhaps, too, Trump's intuitive numerical appeals to the former glories of the economy and the resurgent job creation rate worked, in part because liberal calls for further lockdowns seemed to replicate the kind of sanctimony-laden elite bias to Democratic economic policy that Newfield's chapter explores. But Trump lost, his staff reduced to endlessly contesting the final vote totals in order to redeem their leader's narcissistic attachment to plebiscitary numbers. In Britain, Cummings is out, the prime minister beleaguered, the opportunity to use the public health emergency to kick-start digital governance frittered away.

The old mainstream politics stoked destructive disaffection with the unmet promise of a rational, competent quantocracy. Populism is perhaps even more vulnerable to such crises of public trust, given its emphasis on volatile indicators of popularity and the intense expectations of its voter base. And in the end, the downfall of Donald Trump and the disgrace of Boris Johnson were accelerated by another form of intuitive quantification, the triumph of a neat and addictive visualization that also concealed its relationship with expertise: the COVID dashboard, its mercurially fluctuating case numbers juxtaposed with the remorselessly rising toll of American and British deaths.

1.5. CONCLUSION

Citing years of frustration over their advice being misunderstood, misrepresented or simply ignored, America's foremost experts in every field collectively tendered their resignation Monday.

—*The Onion*, June 16, 1999

The populist triumphs of 2016—Brexit and the election of Donald J. Trump—have been considered a backlash against technocrats and the quantitative evidence they purvey. The sudden fragility of quantocracy unsettles the conventional wisdom that quantitative experts dominated the policy process in the late twentieth and early twenty-first centuries, their apparently objective numbers displacing much of policymaking from the realm of "politics." This analysis argues that the depoliticizing power of quantocracy was always exaggerated. Well before 2016, the relations between quantification, expertise, and depoliticization were beginning to fray.

The heroic ideal of quantocracy promised governance according to "what works": competent, rational, and objective. The disconnect between this ideal and the reality would help to create a crisis of trust in the mainstream system of expertise within politics. The modern expansion of quantitative tools of governance has often been less the consolidator of expert power than an expression of its weakness. Numerical performance indicators and public opinion measurement were introduced in response to right-wing attacks on the efficiency and integrity of the mainstream policy system. They aimed not to grant experts free rein but to reduce their room for discretion. These forms of quantification against experts presaged populist hostility. Rather than increasing public confidence, quantification became increasingly associated with hypocrisy and irrationality. It thus failed as a strategy of depoliticization, as delegation to expert quantifiers itself became a political issue. The result was a paradox: numbers are more ubiquitous and influential than ever in public life, most recently in the form of the big-data revolution, and yet "faith in the technocrat is flagging" (Kelly and McGoey 2018, 1).

At least to some voters, populists appeared to offer solutions to the crisis of quantocratic credibility in the form of a more authentic politics. Against stereotype, numbers have played an important role in the Anglo-American populist idiom. Rejecting the necessity of expert mediation, populist plebiscitary numbers purport to directly reflect the popular will. Nonetheless, these populist leaders did not tackle the underlying crisis of public trust. Indeed, their solutions only intensified the pressure of impossible expectations on policy delivery.

The two trends documented here—the breakdown of the link between quantocracy and depoliticization, and the rise of numbers without experts—are not merely the recent creations of Anglo-American populists. They require us to rethink our interpretations of numbers in public life. Our new sociologies of quantification must take into account the asymmetric and expedient nature of the relationship between politicians and experts; the former, and not the latter, often power the most visible numbers in political discourse. We must remember the classic insight that the power of public numbers is neither static nor inevitable. Trust in numbers is contingent: it can rise and fall, its relationship with the cultural status of experts changes over time, and it has unexpected effects. Forging the status and legitimacy of expert quantification was a long process. Its unmaking and remaking will be equally complex and contested.

NOTES

1. "Populism" is among the most hotly contested terms in contemporary politics. Here I rely on Cas Mudde's influential definition: populism is "an ideology that considers society to be ultimately separated into two homogeneous and antagonistic groups, 'the pure people' versus 'the corrupt elite', and which argues that politics should be an expression of the *volonté générale* (general will) of the people" (Mudde 2004, 543).

2. Vote Leave advocate and Labour Member of Parliament Gisela Stuart, quoted in Witte (2016).

3. Using the searchable database www.thetrumparchive.com, the survey examined the period between Trump's declaration of his presidential candidacy on June 16, 2015, and May 28, 2017, thereby covering his time both as a candidate and as president. This gave a set of 8,825 tweets (including retweets and deleted tweets).

4. For further discussion of this scholarship and its "Original Critique" of quantification, see the introduction to this volume. The development of double-entry accounting practices was a crucial, parallel trend (Poovey 1998).

5. On such struggles for technocratic autonomy and how reliant assertive agencies are on coalition building within the wider political system, see Carpenter (2001).

6. The American public's estimate of the average number of cents wasted per dollar of federal spending rose from 38 cents in 1986 to 51 cents in 2011, for example.

7. See Stephen John (chapter 6 of this volume) for a provocative analysis that spurious numbers are both ubiquitous and surprisingly useful for public discourse and education.

8. This excludes four tweets mentioning presidential adviser Kellyanne Conway, whose Twitter handle is @KellyannePolls, but which do not otherwise refer to polling. The figure would be even higher if we include retweeted graphics or comparisons with other underestimated candidates (invariably Reagan).

9. They often occur together; such cases have been excluded.

10. Beloved Republican phrase "the silent majority" had a brief surge in the early months of his campaign—never mind that the phrase is most associated with Richard Nixon, a president with whom Trump is now compared for quite different reasons.

11. A similar logic applies to another Trumpian metric, retail sales: "2019 HOLIDAY RETAIL SALES WERE UP 3.4% FROM LAST YEAR, THE BIGGEST NUMBER IN U.S. HISTORY. CONGRATULATIONS AMERICA!" (Twitter, Christmas Day, 2019).

CHAPTER 2

The Role of the Numerical in the Decline of Expertise

Christopher Newfield

The research behind this volume was funded toward the end of Barack Obama's presidency and completed toward the end of Donald Trump's. The transition from one to the other made the politics of the numerical a national issue. Obama embodied the rule of apparently dispassionate expertise, while Trump embodied a relentless attack on it. During one of the 2016 presidential debates between Hillary Clinton and Donald Trump, when the candidates were asked about trade deals that allowed effects like Chinese steel imports replacing US steel, Clinton pointed out that Trump had used Chinese steel in his hotels and thus given jobs to Chinese, not American, steelworkers. Trump responded, "She's been doing this for 30 years; why the hell didn't you [fix] it over the last 15 or 20 years? . . . You do have experience. The one thing you have over me is experience. But it's bad experience" (Golshan 2016). Trump agreed that Clinton had massive expertise, but claimed that her expertise had made a mess of everything.

In this context, it seems that Trump, his Republican Party, and the fake newsers in his mass base must be blamed for the decline of expertise. The victims include the horrified professional middle classes who have long voted for the Democratic use of numbers to grasp larger economic and social patterns, such as racialized unemployment, so that they can be addressed (or politely finessed). In familiar contrast, Trump brought a long history of fabrication to the White House, including the "birther" claim— that Obama was born in Kenya. Once in the White House, his lying set some kind of world record for a head of state: "30,573 false or misleading claims," according to the *Washington Post* fact checker (Kessler 2021).

Though Trump and his base came to dominate public discussion of the fate of expert knowledge during his presidency, they are only part of a larger story. In this chapter, I aim to tell another part. I will analyze the role and effects of the expert use of expertise, illustrated by the Barack Obama administration. We are used to complaints about the limits of technocratic governance, both in its alliance with neoliberalism and in the failure of an anti-intellectual culture to appreciate it. Here I'll focus on the Democrats' use of expertise to weaken the wider public's sense of their own historical agency in relation to an economy and society that experts end up defining in fatalist quantitative terms.

2.1. DEMOCRATIC NUMERICAL CULTURE?

The start of the 2020s will be remembered for the SARS-CoV-2 (COVID-19) global pandemic and, in the United States, for the long crescendo of the president's public war on fact, which inspired strangely related events, such as the armed militia protest of public health mandates like mask wearing in the Michigan State Capitol on May 1, and the neo-Confederate assault on the Capitol to block the certification of the Democratic presidential victory six days into 2021. These and other events arose largely in response to falsehood propounded in the highest places: that COVID-19 was just the flu, that mask mandates were a liberal attack on liberty; that a Republican presidential victory had been stolen by massive electoral fraud. There were countless charges and grievances, and yet the events of 2020 seemed to offer a helpful simplification about the motives of Donald Trump's supporters. While the 2016 election had produced a sustained debate about how to rank the apparently multiple issues driving his MAGA base (named for his campaign slogan, "Make America Great Again"), the 2020 election, and Trump's denial of its result, tended to reduce these to two: authoritarianism and white supremacist views, whether formal or informal. Other motives, like "economic anxiety" (Global Strategy Group and Yang 2017) or concern about Washington corruption ("drain the swamp"), faded into the background.

I agree on this ranking of MAGA motives, in part because I've long held that the interaction of racism and authoritarianism shapes the core features of US democracy (Newfield 1996). But identifying constitutive features doesn't explain why these features became more important at various times or why one political demagogue gathers a mass base while others do not. Trump's base was not reduced but nourished by his manifest failures to fulfill his campaign promises or manage COVID-19 "like a successful businessman." Racism and authoritarianism don't in themselves explain the affective grievance culture that animated them—the sense in the MAGA base that their natural legitimacy and centrality had been stolen from them.

I will assert here that the major catalyst with the MAGA base was its sense of victimization—specifically by the system supposedly built by Democrats. After the 2016 election, analysts tried to identify the particular socioeconomic forms of victimization that explained Trump's success, such as unemployment, which he had blamed on "bad trade deals" or "illegal immigration." These efforts were largely unsuccessful. The grievance we frequently heard was that MAGA views were despised and excluded by liberals, elites, experts, and the mainstream media. Were the United States the land of liberty and fair democracy that it was supposed to be, MAGA views would be important, dominant, and plainly respected. My starting point here is my sense that white supremacist and authoritarian sentiments were catalyzed by an epistemic grievance against liberalism that adherents attributed to Democrats. The MAGA base rebelled not only against the content of liberal ideas, like immigration reform, affirmative action, or a woman's right to choose abortion, but against the practices of evidence and argument that give liberal Democratic ideas their social centrality.

This root of MAGA grievance culture—the accepted epistemic validity of mainstream liberalism—brings us to the topic of this volume. We stated in our Introduction that the Original Critique of quantification identifies problems of both knowledge and power. Quantification tends to strip out history, context, and local knowledges, thus partially falsifying any particular situation. Quantification also weakens the power of people in particular situations either to govern themselves or to make others care about their perspectives and experiences, since governing authority easily slips from that particular situation to some other, greater authority elsewhere—a supervisor, a management team, a chief executive, a rating agency, a funding body, and so on. Numbers are both *reductive* and *managerial*. They avoid these effects only with deliberate effort.

This volume both extends and challenges this critique. In chapter 1, Elizabeth Chatterjee observes that critics of the reign of the numerical, though rightly concerned about a "democratic deficit," too easily assume "that depoliticization through quantification is typically *successful*." She goes on to show the extent to which the US and UK political systems have marshaled numbers for political ends and also the involvement of experts—the "quantocracy"—in this marshaling. Leaders of the political right, however deeply anti-intellectual their populist stance, have developed what she terms "plebiscitary numbers"—often popularity ratings supposedly confirming their mass support. Furthermore, Chatterjee outlines a genealogy of the right's politicized use of numbers that traces it back to its political opponents, New Labour and the New Democrats, whose leaders, including Tony Blair, Bill Clinton, Al Gore, and Hillary Clinton, embraced the quantocracy for political gain. She thus deploys the Original Critique's claim that quantification can damage democracy while extending it to conclude that this happens as readily through the politicization as through the depoliticization of numbers—and that it can be performed as readily by centrists affirming experts as by right-wingers lambasting them and the fake news they produce.

In this chapter, I join Chatterjee in focusing on the numerical sins of proexpert politicians rather than those of their anti-intellectual detractors. This means the US Democratic party in the Clinton-Obama quarter-century. During the Democrats' prior fifty-year heyday from around 1930 to 1980, the party had two issues on which it based its mass appeal: (white) egalitarian economic development and, later, civil rights and racial justice, built also on feminist and queer insurgencies over many decades. These achievements were limited and the issues never faded in public importance. On the contrary, they gained relevance as Republicans became more militant about rolling back any progress made toward economic and racial equality, emboldened by the presidency of Ronald Reagan in the 1980s. And yet these were precisely the issues that the New Democrats downplayed, marginalized, or even opposed. After Reagan walked over prolabor and late New Dealer Walter Mondale in the 1984 presidential election, the Democrats ran technocrat Michael Dukakis, who could not summon the civil rights language to defend his candidacy against George H. W. Bush's racial dog-whistling. (For example, Bush's Willie Horton campaign ad implied both that Black men were inclined toward rape and that civil-rights Democrats were enabling them.) In the next election, Bill Clinton acted as though Dukakis hadn't backpedaled either equality agenda quickly enough. His 1992 campaign featured "tough on Blacks" racial signals like the execution of a mentally incapable Black prisoner in Arkansas, Ricky Ray Rector, and a gratuitous attack on radical Black musician Sister Souljah. This continued once Clinton took office, through his abandonment of his own nominee to head the civil rights division of the Department of Justice, the Black legal academic Lani Guinier; his promotion of federal crime legislation that increased incarceration rates for Black and Latinx communities and for which Joe Biden had finally to apologize in 2020; the downsizing of welfare through his critique of welfare "dependency"; and so on (Applebome 1992; Clinton 1996; Lauter 1993; Phillips 1992; Ray and Galston 2020).

Overall, the Democratic retreat from economic and racial equality worked politically for two exceptional Democratic politicians—Bill Clinton and Barack Obama—but failed to build Democratic majorities in Congress, most state legislatures and other local jurisdictions. Donald Trump's victory in 2016 built on Republican control of both chambers of Congress and most statehouses. Republican power, used to oppose both types of equality and then emceed so masterfully by the offensive and divisive Trump, intensified the demands of the Democratic base to build forward on the basis of the double-equality agenda—for starters—that had been neglected for about forty years.

As we made final changes to this manuscript, Joe Biden had replaced Trump in the White House, but had to work with a tied Senate, a smaller majority in the House, and Republicans controlling nearly two-thirds of statehouse chambers. Holding the spotlight on the Democratic use of numbers may be particularly useful in the 2020s. The spectacular toxicity of Trump's final year amped up long-standing concern about his followers' apparent contempt for facts and thinking, a contempt Trump both celebrated and capitalized on: "I love the poorly educated," he had remarked after a string of 2016 primary victories (Stableford 2016). But that is also a distraction from the political and epistemic weaknesses of the Democrats' modes of address.

Multitudes of voters across the spectrum want to be heard by their political leaders, don't think they are being heard, and don't like the various substitutes for being heard that are in common use. One of these substitutes is the collection and use of numbers to override everyday experience. The numerical can be deployed to discredit or evade democratic deliberation and the presence of lived experience in that deliberation. Using expert knowledge against experience, especially dispassionate quantified knowledge and its claim to superiority, undermines the Democrats' historic equality agenda, as I will show. Whether or not the party leadership explicitly moves in the 2020s toward equality, it's worth seeing how its numerical culture, so trustful of large bodies of information and rooted in professional-managerial expertise, will need to be changed for such a shift to endure.

A quick terminological note: most simply, *quantification* refers to descriptions of things or relations between them in the form of numbers; in contrast, qualitative, or verbal, descriptions identify features or characteristics without numbers. The numerical has two important features that the verbal in itself does not. First, it is a mode of comparison that lends itself readily to ranking (Fourcade 2016). Second, it claims objectivity. To generalize on the contrasting term, most people see qualitative description as expressive or subjective. Even when a qualitative assessment seems precise and verifiable, it remains tied to particular perceptions, contexts, or people. The numerical generally has more authority than the qualitative, given the conventional wisdom that it is more objective. The Original Critique identified the tendency of authorities to use numerical discourses, intentionally or not, to correct and exclude vernacular perspectives, rather than to explore and enhance them. This issue has not gone away.

2.2. RACIAL AUTHORITARIANISM IN AMERICA

Trump's 2016 Electoral College victory produced a wild scramble to explain his voters. Gender bias seemed to play a role, though its explanatory power was blunted by Trump's success with the white female vote, in spite of a long series of complaints of sexual harassment and assault, and his recorded boast of enjoying routine sexual groping of women "if they're beautiful" (BBC Staff 2016). Perhaps the bias at work would be better understood as antifeminist.

The leading early thesis was that the white working class turned to Trump because they felt "economic anxiety" and had confidence that the smart, successful businessman could fix trade deals and the rest of the economy (Barabak and Duara 2016). This thesis did not survive detailed analysis of polling data, since Trump's support was much stronger among whites making more than $50,000 per year than among those making less. It quickly became clear that Trump was largely a white middle-class phenomenon (Confessore and Cohn 2016).

The most durable explanation for the Trump vote has been white racism. Trump "won the white vote nationally by 21 points (one point more than Romney), and his campaign rallies were Woodstocks for bigots" (M. Davis 2017). A consensus emerged that racism fed the Trump vote (Clement 2017).[1] As one pollster, Nick Gourevich, put it to Thomas Edsall,

> within economically distressed communities, the individuals who found Trump appealing (or who left Obama for Trump) were the ones where the cultural and racial piece was a strong part of the reason why they went in that direction. So I guess my take is that it's probably not economics alone that did it. Nor is it racism/cultural alienation alone that did it. It's probably that mixture. (Edsall 2017)

Trump voters were more likely than others to tie economic problems to excessive immigration. They were more likely to rail against "illegals" and to want to build a wall along the Mexican border. They were more likely than other white voters to view everything, including economic problems, through their fear of "cultural displacement" (Cox et al. 2017).

Trump systematically stoked this race-based resentment. He did not dog-whistle racist tropes but shouted them out in the open. He spent much of 2020 suggesting that Black Lives Matter protests rested on no legitimate grievance but were borderline terrorist threats to the sanctity of the suburbs. He invited two white homeowners who had brandished assault rifles at peaceful marchers to the Republican National Convention. In addition, in July 2018, his Department of Education's Office of Civil Rights had affirmed white grievance culture by rescinding an Obama administration guidance that allowed restricted use of race as a factor in school enrollments or university admissions (Green et al. 2018).[2] Consideration of candidates' race in enrollments or admissions, or "affirmative action," is not a sideshow in racial politics, but a charged example of what its opponents call "reverse discrimination" against whites. These opponents are still in the majority: in one 2017 poll, 55 percent of whites said "they believe there is discrimination against white people in America today" (Gonyea 2017). Polling both before and after the 2016 election found that Trump supporters were more likely than other whites to agree that "whites losing out because of preferences for blacks and Hispanics" was a bigger problem than preference for whites. "The odds that a person who feels strongly that whites are losing out supports Trump are more than three times higher than for a demographically and financially similar person who feels blacks or Hispanics are losing out or that neither group is losing out more" (Clement 2017). Other studies also suggested that whites in similar economic and educational circumstances would break for or against Trump depending on whether they were for or against crackdowns on immigration, Black Lives Matter protests, and the like, with white racial victimization as the common thread (Craig and Richeson 2014; McElwee and McDaniel 2017).

In addition to income, higher education was a reliable breakpoint for Trump voting, with Trump favored more by people who had not gone to college than college graduates of similar incomes. The statistician Nate Silver pinpointed the "college divide" early on:

> I took a list of all 981 U.S. counties with 50,000 or more people and sorted it by the share of the population that had completed at least a four-year college degree. Hillary Clinton improved on President Obama's 2012 performance in 48 of the country's 50 most-well-educated counties. And on average, she improved on Obama's margin of victory in these countries by almost 9 percentage points, even though Obama had done pretty well in them to begin with. (Silver 2016)

Silver found the reverse in the counties with the lowest percentage of college graduates:

> Clinton lost ground relative to Obama in 47 of the 50 counties—she did an average of 11 percentage points worse, in fact. These are really the places that won Donald Trump the presidency, especially given that a fair number of them are in swing states such as Ohio and North Carolina. (Silver 2016)

Nor was this due to that difference in income which is correlated with higher education:

> I identified 22 counties where at least 35 percent of the population has bachelor's degrees but the median household income is less than $50,000 and at least 50 percent of the population is non-Hispanic white . . . Clinton improved on Obama's performance in 18 of the 22 counties, by an average of about 4 percentage points. (Silver 2016)

Through a series of such comparisons, Silver concluded that lower education was the more important driver of the white Trump vote than higher income—though the latter also helped. In short, the MAGA base came to be codified as the white middle class that did not finish college and read all events through a racial anxiety that this base never subjected to scrutiny.[3] By the time Trump became Super-Trump in 2020 and was winding down his "stop the steal" blowout, the convergence of visible white nationalism and authoritarian insurgency seemed to confirm the 2016–17 view that the racism of the poorly educated was the Trumpian rocket fuel.

It might seem at this point that our main analytical work is done. In our reading so far, the MAGA base liked Trump's promise to restore an America in which they were economically and culturally comfortable and also preeminent by virtue of their white race (Devega 2017). Furthermore, in MAGA America, their security was not dependent on college achievement or on any related cultural or linguistic skill, for they had no obligation to acknowledge, respect, or get along with different kinds of Americans. Nor did they need to compete with poorly paid foreigners for middle-class jobs. Women in MAGA America also would not constitute a threat of difference, or present the need to rethink anything. Women would not have abortion rights, equal pay, or experience the various benefits sought by the several waves of the feminist movement. There were no interpretive or epistemic challenges in the coming MAGA America: a hermeneutics of otherness would never be required.

This view was mostly in place by the end of 2017, and helped rally strong Democratic turnout in the 2018 elections, since this MAGA model turned the Republican base into the implacable enemy of government, affordable health care, equality, climate solutions, and the very presence of people of color. When Trump dismissed and thereby aggravated the two biggest issues of 2020, the COVID-19 public health crisis and anti-Black police racism, he reinforced the model. As 2020 Trumpism intensified the neo-Confederate ethos, the MAGA model of the noncollege white racism base was locked into place.

The 2020 election did not change this analysis of Trump's white patriarchal appeal so much as it intensified and psychologized it. Trump increased his popular vote total from 63 to 74.2 million. As an incumbent president who had dominated the media every day of his four-year mandate, he had eliminated any shred of uncertainty about what a Trump voter was voting for.[4] In 2020, Trump voters pulled the lever with eyes wide open. This fact has fed analyses of Trumpism as a mass psychological condition that pushes racism and inadequate education past previous bounds.

Discussion of MAGA authoritarianism had been building through Trump's term (Edsall 2018; Krugman 2017; Toscano 2017). Interesting examples of book-length treatments appeared during the 2020 election season. Two in particular reinforced each other with complementary methods. In *Authoritarian Nightmare*, former Nixon counsel John Dean partnered with social psychologist Bob Altemeyer to tie Trump partisans to high scores on an instrument that generates a Right-Wing Authoritarian (RWA) scale, which has been in use since the 1970s. They were able to combine this scale with a Monmouth University Polling Institute survey of Trump voters. Their conclusions cast Trump voting as a symptom of cognitive impairment with a fixed authoritarian personality structure. Here is their remarkable summation:

> The verdicts are in. (1) Donald Trump's supporters are, as a group, highly authoritarian compared to most Americans. (2) They are also highly prejudiced compared to most Americans. (3) You can explain the prejudice in Trump's supporters almost entirely by their authoritarianism. (4) Authoritarianism is a strongly organized set of attitudes in America that will prove very difficult to reduce and control. . . . The pillars of Trump's base, white evangelicals and white undereducated males are highly authoritarian and prejudiced. . . . The connections among prejudice, authoritarianism and support for Donald Trump are so strong that no other independent factor can be as important in

supporting his reelection. There just is not much left to be explained, which is a highly unusual situation in the social sciences, but that is where the data have taken us. Ask a very complicated question: Who are Trump's staunch supporters? Get a very simple answer: Prejudiced authoritarians, and a few others. (Dean and Altemeyer 2020, 224–25)

These extremely confident pronouncements conflict with evidence of Trump-voter diversity (Global Strategy Group and Yang 2017; Olsen 2021). They are being applied to around 74 million Trump voters. They also hew so closely to hostile stereotypes of Trump supporters that I was uncomfortable crediting them. And yet they did fit with my own perception of Trumpist discourse in the context of my own research on intractable authoritarian elements in American culture. I was also persuaded by the combination of Altemeyer's empirical work in conjunction with abundant evidence in MAGA circles of a key cognitive feature of authoritarian thinking—the preference for authority over experience. This was the foundation of the widely noted denialism of verifiable evidence such as Joe Biden's electoral victory in November 2020. There is also growing evidence for the label of racist authoritarianism from MAGA ethnography (Hoffman 2020).

Further reinforcing this picture of an authoritarian cognitive death are the psychologists, both those who consider Trump clinically sociopathic and those who focus on his followers. The psychologist Steven Hassan argues that Trump has deployed the standard programming techniques that cult leaders always use to acquire and mentally dominate their followers. He concludes that members of the MAGA base should be treated sympathetically, as victims of techniques of control who must be led carefully back to "their true, or authentic selves" (Hassan 2020).

It may appear that we've lived through the following evolution: white middle-class racism, mostly of the noncollege kind, put Trump in the White House in 2016. There he used his media and propaganda apparatuses to transform racial ideology into a psychological paralysis that was in turn rooted in the underappreciated authoritarian bedrock of US culture (Newfield 1996). Hostile judgments, however, will only drive the MAGA base into deeper states of rage and dissociation, so therapeutic sympathy should be deployed before the country breaks in two. This was the tacit theory behind Joe Biden's decency campaign in 2020: Trump was bad but his followers were not; once given a reasonable choice, they would return to good, caring, highly professional government that Biden was in early 2021 systematically putting into practice, and be naturally attracted to its practical success. The MAGA base had dealt with economic anxiety by transforming it into racist projection. Perhaps expert competence that reduces economic deprivation and injustice would allow the base to dial its racism back. Republican majorities also like the US Postal Service, Medicare, and higher taxes on the wealthy, and the achievements of the Democrat's pros might stand them down.

Here we reencounter our question of experts and their effects on contemporary US political culture. What aspects of professional expertise could really appeal to those attracted to racist authoritarian modes of US life?

2.3. NUMERICAL FATALISM

Earlier I set aside economic factors in Trump's 2016 success. Now I'll reintroduce them in a specific form, as a deprivation of personal agency justified by invoking quantification. I put this in a context in which racism and capitalism don't compete for influence but interact continuously, as the concept of "racial capitalism" encourages us to see (Robinson 2007).

Pollsters found that Trump voters who earned middle-class salaries *were* anxious economically about threats to their standard of living. They also—and this is the key point—did not think the Democrats would help them. In one study, the group that switched from Obama to Trump held a crucial belief: congressional Democrats were even *more* likely to favor the rich than congressional Republicans, and so they put their faith in the supposed egalitarianism of Donald J. Trump (Carnes and Lupu 2017).

This is a startling finding. Slightly more Obama-Trump voters see congressional Democrats rather than Republicans as the party of the rich; twice as many see Democrats as pro-plutocrat compared to Trump.

The plot is thickened by the fact that many Republicans want to vote against plutocracy: nearly as many Republicans as Democrats in exit polls thought "big business has too much influence in American politics" (Edsall 2017). A study looking at historically Democratic counties that flipped to Trump found that all but one of them had voted for Obama at least once, *and* that in nearly all of them a factory had closed during the campaign season, forming "embittering reminders that the 'Obama boom' was passing them by" (M. Davis 2017). One of the authors of another study put it this way to the *New York Times*'s Edsall: "The biggest common denominator among Obama-Trump voters is a view that the political system is corrupt and doesn't work for people like them" (Edsall 2017). A longtime Democratic pollster, Stanley Greenberg, elaborated on the history:

Working-class Americans pulled back from Democrats in this last period of Democratic governance because of President Obama's insistence on heralding economic progress and the bailout of the irresponsible elites, while ordinary people's incomes crashed and they continued
to struggle financially. . . .

In what may border on campaign malpractice, the Clinton campaign
chose in the closing battle to ignore the economic stress not just of the
working-class women who were still in play, but also of those within
the Democrats' own base, particularly among the minorities, millenni-
als, and unmarried women. (Greenberg 2017)

Trump seems to have made real inroads in many Obama precincts with
his promise to bring jobs back. He did this specifically with people who
had voted Democratic in the past and had felt betrayed by them. In this
logic, having voted for Obama and seen no disruption in policies favor-
ing the rich, they were willing to take a chance on a guy who proclaimed
himself disrupter-in-chief.[5]

Democratic economic policy has been deeply disappointing for people
who care about stability or equality. But it hasn't been worse than Republican economic policy on those points. We are in danger of arriving back
to where we started: Trump picked up those skeptics about Washington
economics who were also more likely to blame immigrants and racial minorities for their problems. Trump, in other words, got the racist doubters of capitalist economics, while Clinton got the less-racist or antiracist
doubters. Didn't we already know this?

Yes, but there's another differentiator between Democrats and Trump's
type of Republicanism. Trump describes economics as a triumph of the
will, American-style. Democrats have been describing economics as the
outcome of quantitative laws.

I'll offer two examples of what I'll dub *numerical fatalism*. The first is
from the cornerstone text of New Democratic political economy, Robert
Reich's 1991 *The Work of Nations*. Reich's book is about how the United
States should fit its labor force into the global economy. It was superbly
timed, appearing as the Soviet Union was collapsing and the global workforce was undergoing the Great Doubling (Freeman 2006), and it helped
earn Reich an appointment from Bill Clinton as his secretary of labor.

Reich made fundamental claims that solidified an enduring New
Democrat common sense. He wrote that global economics was undergoing structural changes that were beyond the reach of national policy. He
wrote that eventually national economies would cease to exist (Reich 1991, 3)
and there was no point in a new administration in Washington trying to

control global economic forces or the offshoring of good union jobs. He wrote that the new firm would be an "enterprise web" and would thus disperse rather than concentrate economic power (88): "Instead of imposing their will over a corporate empire, [executives] guide ideas through the new webs of enterprise" (97). He wrote that power and resources would flow to skill rather than to existing wealth or ownership. He classified levels of skill into three types: routine production workers, in-person servers, and symbolic analysts. He wrote that the first group, who worked on the floor at Ford or US Steel or Carrier, belonged to the past economy; government effort could not save their jobs. He wrote that the second group were tied to a location (home care workers had to be onsite), and were not going to be highly paid but could not be offshored. He wrote that the third group contained the value creators of the new economy. Their services "include all the problem-solving, problem-identifying, and strategic-brokering activities" (177); they manipulate symbols that stand for reality, and their manipulation yields new efficiencies, new financial arrangements, new deployments of resources, new music, new designs, new ad campaigns (178). He wrote that the main job of government is to increase worker skills (and the share of symbolic analysts) by funding training and education. He admitted that the country was showing signs of a "politics of secession," in which "growing segregation by income" (274) would encourage the wealthy to set up "homogeneous enclaves" for themselves (268). But he concluded that this inequality would be managed by a shared "positive economic nationalism" (312) in which the refusal to control trade and protect American jobs would be accompanied by generous public spending to help people raise their own productivity so they could compete and win. "In principle, all of America's routine production workers could become symbolic analysts and let their old jobs drift overseas to developing nations" (247).

All of Reich's key predictions were wrong, including the one about brainworkers being protected from economic turbulence in contrast to the other two categories. Over the next quarter-century, job loss and secession accelerated—deindustrialization and inequality are the hallmarks of the period, even as the economy continued to grow. Wealth and power did not become decentralized through the "network" economy. Routine production workers did not launch themselves onto the career path of symbolic analysis. Public funding did not generously support everyone's transition to a higher level of capability. As important as Reich's inaccuracy was his justification of change. In Reich's model, all this shifting of employment, production, and pay happened more or less automatically. Deindustrialization wasn't the fault of the CEOs and governing boards who made the decisions, since they were just other symbolic analysts in a network processing global economic data and doing what the numbers said they had to do. Deindustrialization wasn't the fault of Roger Smith or his successors at the top of General Motors, or the fault of the Democrats, although they occupied the White House for two-thirds of the 1991–2016 period. It wasn't really even the fault of the Republicans. All Democrats should do was encourage former production workers to adapt to the inevitable and give them retraining or college money in the teeth of Republican opposition. Agency, in the New Democrat model, lay with global structural forces, not with top officials in the private or public sectors. Reich got Democrats off the hook, in their own minds, for passive adaptations to forces that wreaked the most havoc on their own base.

Of course, Republican economic policy was actively rather than passively plutocratic, so Democrats stayed in national contention. Twenty-five years after Reich's book appeared, President Barack Obama held a town hall in Indianapolis, Indiana, on June 1, 2016, in the midst of Hillary Clinton's campaign for president. The moderator, Gwen Ifill, tried to prompt Obama to deal with people's experiences rather than facts and figures. He struggled to do this. He said things like, "We have fewer federal employees today." She said, "um-hmm." He continued,

> If people are feeling insecure and they're offered a simple reason for how they can feel more secure, people are going to be tempted by it, particularly if they're hearing that same story over and over again. The health care costs since I signed Obamacare have actually gone up slower than they were before I signed it. Twenty million more people have health insurance. So the arguments they're making just are not borne out by the facts. (Ifill 2016).

Ifill responded by quoting Bill Clinton back to Obama: "'Millions and millions and millions of people look at that pretty picture of America you painted,' which you just described, 'and they cannot find themselves in it to save their lives.'" You think the people are being seduced by a false story, she was saying, but you are giving them false statistics—true in the abstract, but false to their actual experience. "Why is there a disconnect?" she asked Obama.

He took up again with his tale of fated effects : "Well, look, here's what has changed in the economy over the last 20 to 30 years. . . ." The striking feature of Obama's account is the passive voice: "And what started happening is you started seeing foreign competition. Unions started getting busted." Nobody does things, they just happen. It's the tide of history:

there's no set of bankers, CEOs, think tank free-marketers, anti-union law firms, or Republican senators who did bad things and who now must be stopped.

Ifill gave up and said, "let's turn to the audience and see what they think." The first question was from a "fifth-generation fruit and vegetable grower here in Elkhart County" who was struggling under new paperwork for the Food Safety Modernization Act and Obamacare. How can we encourage younger people to enter this kind of marginal business, he asked. Obama gave a seven-paragraph answer that said he was for regulation and also against outdated regulations and that "some elements of the regulations I put in place have probably put a burden on you"—but it's for the best.

Then he got a question that laid out the problem all over again.

ERIC COTTONHAM: My name is Eric Cottonham and I'm representing the Steelworkers Union, Local 1999. And I'm trying to find out, what do we have left for us—all of our jobs are leaving Indianapolis. I see here you're doing a lot of things, but in Indianapolis, there's nothing there for us. I mean, what's next? I mean, what can we look forward to in the future as far as jobs, employment, whatever? Because all of our jobs have left or in the process of leaving, sir.

PRESIDENT OBAMA: Well, in fact, we've seen more manufacturing jobs created since I've been president than anytime since the 1990s. That's a fact. And you know, if you look at just the auto industry as an example, they've had record sales and they've hired back more people over the last five years than they have for a very long, long time.

We actually make more stuff, have a bigger manufacturing base today than we've had in most of our history. The problems have been— part of the problems have had to do with jobs going overseas and this is one of the reasons why I've been trying to negotiate trade deals to raise wages and environmental standards in other countries, so that they're not undercutting us.

But frankly, part of it has had to do with automation. You go into an auto factory today that used to have 10,000 people and now they've got 1,000 people making the same number of cars or more. And—so what that means is even though we're making the same amount of stuff in our manufacturing sector, we're employing fewer people. . . .

Obama carried on like that for another seven transcript paragraphs. He overrode Cottonham's description of local conditions by invoking the quantitative "fact" of its opposite, a manufacturing boom. This rejection of experience is point blank something people hate about experts, but it was Obama's routine. He never came close to answering the question. He had no suggestions for things people in Indianapolis could do. He limited his own political agenda to supporting workers who adapt to job loss with retraining as opposed to addressing the job crisis directly. He praised clean tech and other new tech jobs, "like 3-D printing, or, you know, nanotechnology," and went on to reiterate that the days when "just being willing to work hard" meant a steady job are over. So, "you cannot look backwards, and that doesn't make folks feel good sometimes . . . but they're going to have to retrain for the jobs of the future, not the jobs of the past."

In short, Reich's fatalism carried on through later Democratic policy: neither governments nor the corporate world can control economic ground rules and must instead stick with helping people adapt. By 2016, this kind of adaptation had a track record of deindustrialization and inequality: the voters who thought Democrats were twice as plutocratic as Trump in the survey I mentioned were wrong on the specific point but right about the drift of Democratic expert policy. The drift floated on a tide of economic numbers.

Facing the stereotype of a party run by and for the "the college kid who tells me how to do my job," Hillary Clinton neither explained the thinking behind key economic policies, nor atoned for bad results (Williams 2017), nor proposed action against economic policies that had produced them. She adjusted her views around the edges of charged economic issues like the North American Free Trade Agreement, the Trans-Pacific Partnership, and other trade agreements that Trump constantly attacked. But she did not admit long-standing Clintonite mistakes in a way that people associate with basic honesty or professional integrity. She wrapped policies in a protective shield of economic science. In May 2016, Clinton told the CBS Sunday program *Face the Nation*, "I do believe in trade. . . . We are five percent of the world's population. We have to trade with the other 95 percent. That has on balance been a net plus for our economy" (Perry 2016). Of course such numbers had nothing to do with the effect of trade on much of the US population or with what kind of trade deals the United States should have had, as Trump pointed out in every speech. Trump voters were wrong if they thought he would act to bring mass prosperity, but they were *not* wrong to doubt that Clinton would. Her party had naturalized factory closures and mass layoffs as though they reflected the quantitative laws of economic globalization. Her allies had taught two generations of nonuniversity workers that the erstwhile party of working people would come up with statistics to explain why they had no work or had gone from building trucks for union wages to unloading trucks at Wal-Mart.

The Democrats' expert politics had three intertwined problems. One—as Chatterjee discusses in chapter 1 of this volume—was its effects: a market-based, tech-focused Democratic party, led by the Clintons and Obama, spent twenty-five years promoting an economic model that did not benefit most nonuniversity Americans and made the others feel dependent on allegedly second-class (nonknowledge) skills. The Democrats asked people to believe that the model would deliver for them in time, because that was the story the numbers told—if and only if they changed themselves into knowledge workers. Finally, they discounted the qualitative, personal, local, community-based experience that said the Democrats' model wasn't working. The Democrats had applied policies favored by professionals and by the finance and tech sectors to working people, had failed to help them, and had justified the policies in numerical terms.

Trump came along and embedded white supremacist politics in a rejection of the Reich-Clinton-Obama globalization enlightenment. Trump allowed the Republican base to do three things at once: reject Democratic policies; avenge themselves on their embodiment, Hillary Clinton; and affirm the only thing that seemed able to stand up to the experts and their culture of quantification, the truth-defying will of Donald Trump. They were voting for a strongman who asserted the personal power to reject the alleged facts, and who did so every day. The Clinton campaign never grasped this attraction of Trump's lying and denial. If Trump could push back Hillary's free-trade "quantocrats," then, it seemed, they could push back the "truth" of deindustrialization.

Joe Biden and Kamala Harris took the White House back from Trump. Will they also democratize expertise in economic policy?

2.4. PROFESSIONALS AND KNOWLEDGE DEMOCRACY

For Democrats, there's only one right answer to that question, since nondemocratic expertise failed to keep Donald Trump out of the White House. To counter him in 2016, Democrats needed to have run against their own record of numbers-driven deindustrialization and weak responses to structural racism. They needed to engage the local knowledges that had survived or even prospered. They still need to do these things in the 2020s, for better and more successful practical politics.

Democrats should explicitly endorse and implement *epistemic parity* between qualitative and quantitative arguments in political life. I would say the same for professionals and researchers more generally, as parity will improve the quality of sociocultural knowledge in ways that are beyond my scope here. Parity would involve subjecting arguments claiming quantitative objectivity to democratic deliberation. The numerical would encounter qualitative retorts from people in the full range of social contexts. The process would denaturalize arguments grounded in dominant discourses like the economics of trade. Different types of arguments—qualitative and quantitative—would meet as relative equals. General principles and historical findings about economics and academic performance would come from experts; experiences of results, critiques of the paradigm, and alternative models would come from citizens. Citizens and experts would discuss the extent to which corporate tax policies, police budgets, or college admissions standards are political choices rather than laws of financial flows or biological performance. Experts would take vernacular arguments seriously rather than seeing them as methodologically inadequate or backward or in need of immediate nudging (Kuttner 2014). I posit one likely result as a stronger sense of political agency in political subjects, who would then be more regularly engaged with experts.

To illustrate this point, I'll use this principle of epistemic parity to rewrite that exchange between Obama and Cottonham in Indianapolis in June 2016. I pick up where Cottonham concludes, "All of our jobs have left or in the process of leaving, sir," and try to keep Obama's voice in constructing an alternative reply that has been *disquantified* (see Greg Lusk, chapter 9 of this volume), with quantitative and qualitative knowledge interacting across equal footing. Some elements:

1. *Acknowledgment of epistemic gaps, expert blindness, and validation of local, qualitative knowledge.* "You, sir, know quite a bit more than I do about jobs in Indianapolis. I'm very sorry to learn that your community has been struggling like this. We do a lot of big-picture talking in the White House, but it sometimes doesn't fit with the lived experience of workers and communities in the country. What you've told me in your question is essential information, and I will not forget it."

2. *Situating his own expert knowledge as important but incomplete.* "When you mention job loss, I am sorely tempted to launch into one of my trademark lectures about the knowledge economy and globalization. At some point I would annoy you by using my standard line, 'a lot of those jobs aren't coming back.' I'm still a professor, so I have to tell you a couple of things right now about how hard it is to keep our manufacturing base in the teeth of global competition, but I know it's not the whole story."

3. *Invoking his political agency in the company of theirs.* "So that's the big picture. But it doesn't tell us that all we can do is accept and

adapt. That's where government comes in. As Democrats, we see government as an instrument of collective agency—yours and mine and all of ours. We deliberate together in our different modes, we hash it out to minimize a zero-sum winners vs. losers competition, and we use the government's resources to change the course of markets and economies and race relations when these things hurt our people. We don't accept collateral damage in Indianapolis or anywhere else."

4. *Bad agents at whom we can point fingers.* "Many of our efforts to recover from the financial crisis were shrunk or blocked by the Republicans. As you know, their mission on earth is to discredit government so the spoils are given to their sponsors in big business. If we help you, their ideas look bad, so they fight us. When they do, they fight your welfare too. Here in Indiana you have the same problem. Your governor, Mike Pence, has been fighting your ability to get federal support for health care. So help me put them back in the bottle—and I say that to Republicans as much as Democrats: don't let selfish Republican leaders hurt all of you. Vote them out."

5. *Concrete remedies, to be improved by bringing disagreements to the surface and creating expert-grassroots dialogue.* "What do you in Indianapolis think would help keep jobs here? You've tried wage and benefits concessions. They aren't fair, and they don't work to keep companies at home. Here's my idea, one we've been talking about in my administration: tax penalties for offshoring companies. Gwen mentioned you've worked for Carrier. What if we told them, 'You can save money by shipping Indiana jobs overseas, but we'll calculate the difference, and make sure your taxes go up by the same amount.' Trump wants to bully Carrier out of moving—and I agree with him on the goal. But cutting the tax benefits of ending your job is a better way of keeping your job here. Now, there are problems with that idea, and some of my advisers don't like it at all. But I'd be interested in what your Steelworkers local thinks—I'll certainly be hearing from Carrier executives about it—so let's set up a way of getting your views into our White House analyses."

6. *Regular people on a journey (including Barack the citizen).* "You've been through a lot in Indianapolis over the decades. That sounds like a political homily, and it is, but it's true. You've faced a lot of problems here and some have beaten you, but you've solved others. You've done that through your own versions of democratic deliberation and political action. You know I've been a professor, but also a community organizer, and I have an idea of how that process works.

We have data and knowledge at the federal level that can help, and you have knowledge here. I know that you've fought specific job losses successfully, and also fought off Governor Pence's view that his religious-freedom law would allow discrimination against LGTBQ folks (Smith 2015). You struggle against the odds, you are sometimes humiliated and crushed, like I was in the midterm elections of 2010, you come back and gain ground and then lose ground, it can be horrible, but you often wind up winning something important. Your strength and also your knowledge will get you through this."

7. *We fight for you.* "You'll get through this because that's what the Democratic party is about. We'll offer policy and also financial support. Together we take expert knowledge and turn it into political *agency* that makes things better."

That last sentence gums things up with our key terms, but it's on the right track. Experts will experience less political backlash and better intellectual results if they develop parity and reconnection between quantitative and qualitative discourses. This will improve relations between the experts and citizens that epistemic inequality keep apart. Epistemic parity, and the attending sense of power through knowledge, would help to re-democratize everyday political life.

NOTES

1. Analysis Interpretation of the news based on evidence, including data, as well as anticipating how events might unfold based on the past.

2. The rescinded Obama policy (US Departments of Justice and Education 2011) is a useful summary of the pre-Trump limited consensus on affirmative action.

3. The press continued to feature the white middle-class Trump voter (e.g., Carnes and Lupu 2017; Sasson 2016).

4. For example, "Nearly half of the voters have seen Trump in all of his splendor—his infantile tirades, his disastrous and lethal policies, his contempt for democracy in all its forms—and they decided that they wanted more of it. His voters can no longer hide behind excuses about the corruption of Hillary Clinton or their willingness to take a chance on an unproven political novice. They cannot feign ignorance about how Trump would rule. They know, and they have embraced him" (Nichols 2020).

5. The historian Mike Davis (2017) offers a helpful example from one part of Trumpland:

The largest concentration of white poverty in North America, the Southern mountains, have been orphaned not just in Washington but also in Frankfort, Nashville, Charlestown, and Raleigh where coal lobbyists and big power companies have always dictated legislative priorities. Traditionally their henchmen were county Democratic machines and the blue faded from Appalachia only reluctantly at first. Carter won 68 percent of the vote in the region and Clinton 47 percent in 1996. . . . The United Mine Workers and Steelworkers, under the best leadership in decades, fought desperately in the 1990s and 2000s for a major political initiative to defend industrial and mining jobs in the region but were turned away at the door by the Democratic Leadership Council and the ascendant New York/California congressional leadership. Ironically, Clinton this time around did have a plan for the coal counties, although it was buried in the fine print of her website and poorly publicized. She advocated important safeguards for worker health benefits tied to failing coal companies and proposed federal aid to offset the fiscal crisis of the region's schools. Otherwise her program was conventional boilerplate: tax credits for new investment, boutique programs to encourage local entrepreneurship, and subsidies for the cleanup and conversion of mining land into business sites (Google data centers were mentioned—talk about cargo cults). But there was no major jobs program or public-health initiative to deal with the region's devastating opiate pandemic.

Can Narrative Fix Numbers?

CHAPTER 3

Audit Narratives

Making Higher Education Manageable in
Learning Assessment Discourse

Heather Steffen

Since the 1980s, learning outcomes assessment (LOA) has become the dominant mode of understanding and evaluating undergraduate education at colleges and universities in the United States. LOA seeks to evaluate educational quality by measuring students' achievement of specific learning outcomes: skills and knowledge that students should be able to demonstrate upon finishing a course of study. LOA works at the level of the course, program, or institution, and in most instances it does not include evaluating the performance of individual teachers or students. Unlike grades or student evaluations of teaching, LOA produces aggregate measures meant to guide educational improvement and demonstrate universities' accountability to their students, funders, and publics. Examples of influential learning outcomes assessment tools include the Collegiate Learning Assessment (CLA and CLA+) developed by the Council for Aid to Education; the Degree Qualifications Profile developed by the Lumina Foundation; the Measuring College Learning Project; and the Association of American Colleges and Universities' Liberal Education and America's Promise (LEAP) project. In the United Kingdom, learning assessment is found in the Teaching Excellence Framework.

From its beginning, the American assessment movement has continually struggled with a core tension between the "improvement paradigm" and "accountability paradigm" (Ewell 2008). In the improvement paradigm, LOA is a tool for internal use, providing departments and institutions with data to help them enhance their curricula and courses. Improvement-focused assessments are developed with a local context in mind. Their methods reflect local concerns and their results are typically not reported out. The accountability paradigm, on the other hand, seeks to produce comparable, transparent data for consumption by external stakeholders.

The core paradox of the learning assessment movement results from the disconnect between these two impulses: state and federal policies require assessment for purposes of accountability, and to provide access to the "black box" of higher education, while faculty and administrators pursue assessment for improvement that produces little useful information for outsiders (Campbell 2015). This paradox has sparked intense debate between critics of assessment-as-accountability, who view LOA as a method of surveillance and mode of governance, and proponents of assessment-for-improvement, who downplay reporting requirements and emphasize the benefits LOA can provide to students and disciplines.

The thirty-five-year-long debate about learning outcomes assessment in US higher education, and the fact that LOA remains an incomplete, controversial project at many institutions, means that this case of quantitative evaluation can serve as a laboratory for exploring the rhetoric and narratives produced by advocates of social measurement. Social scientists who study quantification examine the history of quantitative reasoning and how it rose to epistemological dominance across social domains, delve into its effects on people and organizations, and analyze the processes by which measures and indicators are negotiated and resisted.[1] But one question scholars have not tackled as fully is: How are regimes of social measurement installed and maintained? In critiques, quantitative rationality can appear to be monolithic and universally successful, when in reality performance measures, metrics, indicators, and targets are more fragile and more at risk of failure than existing accounts of their negotiation, resistance, and effects make clear.

Despite its predominance as a way of knowing, one limit of the numerical is that quantification must constantly argue for itself. Scholars like Michael Power and Wendy Nelson Espeland have drawn attention to the roles of rhetoric and narrative in the maintenance of audit cultures, yet few studies have examined the representational and rhetorical features of the narratives that surround and secure regimes of social measurement.[2] In what follows, I employ literary and cultural studies methods to address several questions that bridge these gaps in the sociology and anthropology of quantification: What kinds of rhetorical strategies are used to install and maintain regimes of social measurement? What roles does narrative play in cultures of audit? What are the representational and rhetorical features of the narratives produced by advocates of accountability, transparency, and audit?

I address these questions by describing what I call "audit narratives"—implicit and explicit stories about audit, auditors, and auditees that appear in the rhetorical performances around programs of social measurement—in arguments produced by learning assessment advocates. In particular, my claims rest on an analysis of white papers commissioned and published by the National Institute for Learning Outcomes Assessment (NILOA), chapters from *Literary Study, Measurement, and the Sublime* (Heiland and Rosenthal 2011), and the 2006 report of Margaret Spellings's Commission on the Future of Higher Education. This corpus offers access to both cross-disciplinary and discipline-specific arguments, as well as an exemplary policy statement. The NILOA papers for the most part do not make disciplinary distinctions and the authors address their arguments primarily to fellow professionals, administrators, and faculty who are already involved in assessment. In contrast, *Literary Study, Measurement, and the Sublime* (*LSMS*) is addressed to scholars in the humanities, especially literary studies, who are skeptical of LOA's ability to measure the full range of experiences and knowledges at work in humanities classrooms. The Spellings Report demonstrates how assessment discourse has been adopted by policymakers at the federal level. In this chapter I examine how the audit narratives surrounding and informing the learning assessment debate define higher education's problems, how they characterize key actors, and how they model agency in higher education in order to present LOA as a commonsense solution.

In contrast to the common assumption that narrative and quantification are opposing modes of communication, this chapter shows how narrative supports the restructuring of qualitative and professional knowledge domains into cultures of audit permeated by quantitative rationality. Section 3.1 details the theory behind my analysis, in particular by describing the concept of audit culture (and its weaknesses) and reviewing how scholars of quantification have conceived the roles of rhetoric and narrative in cultures of audit and accountability. Section 3.2 presents a rhetorical and narrative analysis of the corpus described above. This analysis shows how audit narratives refigure organizational actors as auditors and auditees, establish relations of accountability, and reorient professional values. Audit narratives, I contend, are able to reshape subjectivities, organizational relations, and institutional missions by representing: (1) audit as a universal solution; (2) the auditor as a mediator; and (3) the auditee as both a bad actor and a responsible subject. The chapter concludes with a discussion of how audit narratives in assessment discourse affect undergraduate education in the United States and a consideration of this study's implications for the broader study of quantitative cultures from humanities perspectives.

3.1. NARRATIVE IN THE AUDIT SOCIETY

Published in 1997, Michael Power's *The Audit Society: Rituals of Verification* initiated the study of audit as a key cultural idea, one that drastically exceeds its everyday denotation of a set of technical practices used to check whether an "activity was carried out efficiently and whether it achieved its goals" (3). By the time Power was writing, audit had already achieved hegemonic status as a way of knowing not only financial organizations and supply chains but also social, political, and educational processes, especially in the United Kingdom. Power termed this state of affairs the "audit society," a society "in which a particular style of formalized accountability is the ruling principle" (4). An organizational practice forged in the worlds of finance and quality assurance, audit promises to produce improvement, efficiency, and accountability. Its promises of epistemological certainty are underwritten by the proceduralism of audit practice and the objectivity of auditors. To achieve greater certainty and objectivity than qualitative analysis or expert evaluation can achieve, audit relies on what historian Theodore Porter (1995) has called "trust in numbers," rather than trust in human judgment.

Yet, according to Power, the spread of audit was not driven by its success at providing improvement, efficiency, and accountability. Despite its perceived epistemological power, audit rarely lives up to its promises because its technologies and practices are rarely capable of delivering on the grand, aspirational policy programs to which they are attached. Rather, Power explains, "the growth of auditing is the explosion of an idea, an idea that has become central to a certain style of controlling individuals and which has permeated organizational life." What has spread throughout every domain of the audit society is more than a particular mode of accounting; it is a mode of conceptualizing and knowing social organizations and processes and "a certain set of attitudes or cultural commitments to problem solving" (Power 1997, 4).

For Power and later scholars of the audit society, the implications of understanding audit as a dominant "paradigm for knowledge" are many (Power 1997, 12). Because auditing is not only "a collection of tests and an evidence gathering task" but also "a system of values and goals which are inscribed in the official programmes which demand it" (6), audit has the power to shift institutional and organizational cultures without appearing to do so. Audit's claims to procedural and technical neutrality mask its power to reshape institutional and organizational priorities, goals, and values. In this way, calculative practices can promote antidemocratic management and governance regimes. They "recast political programs as mundane administrative and technical matters to be dealt with by experts," write anthropologists Cris Shore and Susan Wright (2015, 421), "thereby masking their ideological content and removing them from the realm of contestable politics." Shore and Wright's research on what they call "audit culture" explores the effects of audit on public-sector services and organizations, using universities as their case study. When the techniques and programs of audit move "into a new organizational context," they propose, the new context is altered to "mirror the values and priorities embedded within the audit technologies," and new subjects are brought into being (426). In other words, systems of audit and accountability are performative: they perform into being a world in which labor processes, organizational relations, and institutional outcomes are visible, measurable, and subject to accountability.

Scholars have had much to say about the environments and subjects created in audit cultures, but few studies have focused carefully on the means by which audit cultures are able to construct or perform them into being. One exception is Wendy Nelson Espeland and Michael Sauder's intricate ethnographic account of the effects of university rankings on US law schools, *Engines of Anxiety: Academic Rankings, Reputation, and Accountability* (2016). Based on interviews with dozens of prospective law students, professors, deans, admissions officers, and career counselors, Espeland and Sauder define four "mechanisms of reactivity" at work in law schools in the rankings era: (1) commensuration, or the creation of rankings by turning qualitative differences into comparable quantities; (2) self-fulfilling prophecy, or the tendency of highly ranked schools to remain highly ranked due to the rewards that flow to them based on their presumed prestige and success; (3) reverse engineering, wherein schools game rankings systems by tweaking their operations to move up in the standings; and (4) narrative.[3] The first three mechanisms are fairly straightforward, and Espeland and Sauder illustrate them with detailed evidence. But narrative begs more examination as a mechanism of reactivity and performativity in both higher education and the wider study of the audit society.

In Power's *The Audit Society* (1997), references to the interactions between narrative and audit appear occasionally but without theorization. He hints that narrative plays two supporting roles in the audit society. The first is helping to smooth over the expectations gap by reinforcing the programs behind audit's imperfect technologies. "Technical practice cannot be disentangled from the stories which are told of its capability and possibility," he writes, and thus narrative production is "an intermediate and necessary aspect of the programmatic side which connects concrete technical routines to the ideas which give them value" (6–7).[4] Narrative plays a second role in Power's framework by serving to legitimate audit as a way of knowing and governing. It positions audit as central to policymaking and policy decisions by circulating representations of audit's success and potential through a process Power refers to as "meta-accounting." Meta-accounts are stories and rhetoric that give the audit society meaning and promise. While they may be "messy and inconsistent," they are also "performative, projecting and enacting ideals of [audit practices'] capability which legitimate the field of knowledge as a whole" (9). In other words, narrative meta-accounts depict a world in which all problems are solvable via audit, especially if the audit is done well.

A third function of narrative in quantitative cultures is observed in Espeland's essay "Narrating Numbers." She explains that we can, in part, understand the simplification accomplished by quantification as happening through the "erasure of narratives: the systematic removal of the persons, places, and trajectories of the people being evaluated." As commensuration occurs, specifics, causalities, and contexts are left by the wayside. But once indicators and metrics begin to "move about and are re-appropriated or resisted by those being evaluated, they elicit new narratives, new stories about what they mean, how they unfold, if they are fair or unfair, or who made them and why" (Espeland 2015, 56). Quantification erases old stories, and new ones, created in response to the numbers, are told in their place. These new narratives structure our reactions to quantification, allow those who would contest numerical outcomes some agency to do so, and, when told to explain poor performances, can assuage auditees' anxiety about the future and the state of competition. I agree with Espeland that narratives serve these functions, but the evidence shows that they also do much more than this. As Espeland and Sauder write in *Engines of Anxiety* (2016), "quantification is often understood as the opposite of narrative. . . . Yet this binary division of numbers and stories does not capture their complex interplay" (37). The remainder of this chapter explores this complex interplay as it unfolds in the case of learning outcomes assessment.

3.2. AUDIT NARRATIVES IN LEARNING ASSESSMENT DISCOURSE

My hypothesis is that much of the cultural work of audit is achieved not by the actual technologies and practices of audit, but instead by the rhetorical performances that surround and sustain them and by the rhetorical performance of the audit itself. One powerful strategy employed in these performances is the telling of what I call "audit narratives"—stories that explain how audit produces improvement and accountability, and that communicate how actors fit into an audit scheme. The work of narrative in audit cultures is complex, constant, and always incomplete. Audit narratives must make a world and simultaneously propose how to change it. Like most narratives in the policy sphere, they must "create a social reality that suggests it already exists" (Lemke 2001, 203). In doing so, audit narratives inscribe new social relations and circulate new or changed sets of values, priorities, and goals. In Power's terms, audit narratives are meta-accounts that link technologies to policy programs, legitimize audit as a practical solution in a given domain, and define appropriate relations of accountability between participants and other stakeholders.

Audit narratives can be explicit or implicit when they appear in policy and media, as well as in scholarly or gray literature discourses surrounding audit cultures. Explicit audit narratives appear especially in intraprofessional literature circulated between audit experts and in the public genres of advice, promotion, and journalism that report the successes, failures, and results of audits. Implicit audit narratives often work through allusion to existing cultural and media narratives about a domain. After the 2011 publication of Richard Arum and Josipa Roksa's *Academically Adrift: Limited Learning on College Campuses*, for instance, the terms "limited learning" and "disengagement compact" became shorthand for the book's story about the decline of learning in US higher education and its cause: a tacit agreement between students and faculty to lower their expectations and standards. The concept of audit narratives, in other words, comprises not only fully developed stories recounted by auditors, audit advocates, and critics; it also includes the constellations of stories—true and fictional—that are mobilized as rhetorical resources by those involved in debates over audit.

Audit narratives tend to multiply around contested forms and sites of audit. LOA has been a particularly fertile ground for their telling because of the long duration of debate about it, higher education's history of shared governance and faculty control of curricular decision-making, and the diverse set of stakeholders involved in the conversation. My analysis of learning assessment discourse suggests that three major representational and rhetorical features appear in audit narratives: (1) audit as universal solution, (2) auditors as mediators, and (3) auditees as bad actors and responsible subjects.

3.3. AUDIT AS UNIVERSAL SOLUTION

In an analysis of policy narratives in US Department of Education statements, Tatiana Suspitsyna (2010) finds that a primary function of such narratives is to "produce 'obviousness, a common sense' and normality within which the proposed reform measures appear as the most appropriate solutions to identified problems" (578). Audit narratives in learning assessment and other discourses of social measurement perform the cultural work of defining social and organizational problems such that particular forms of audit and social measurement appear as the obvious, sometimes even the only, viable solutions to them. Audit narratives, that is, shape the way we think about social and organizational power structures, who has expertise and who should have agency within them, and how social and organizational change happen. "The idea of audit"—before any particular audit strategy is selected—"shapes the public conceptions of the problems for which it is the solution," Power confirms (1997, 6). For a problem to be solvable by audit, it must be represented as displaying certain traits. Because audit is predominantly a quantitative mode, the problem at hand must include measurable elements, or measurable elements must be created. Because audit is a tool of oversight and accountability, the problem must be defined as resulting from a lack of oversight or transparency, or from failure to act on managers' insights or respond sufficiently to calls for transparency. Audit solves problems that can be managed from above, in which clear goals are established and clear failures to meet them are demonstrated. Auditable problem areas require the identification of discrete groups of auditors and auditees, as well as stakeholders to whom both groups are accountable.

In US higher education, the measurability of individual students' aptitude for learning has been accepted since at least the early twentieth century, when intelligence testing was first invented, and it was quickly reinforced by the introduction and rapid spread of college entrance exams and standardized tests like the Scholastic Assessment Test (SAT). The measurability of a program, major, or institution's learning outcomes is a newer educational concept, but it has similarly become settled knowledge within the educational research community over the last forty years. The long debate about LOA focuses on measurability only at its edges, usually around the question of how to measure "ineffable" skills and traits, like critical thinking, resilience and grit, or aesthetic appreciation. Even here, tests and rubrics like the Collegiate Learning Assessment (CLA+) and California Critical Thinking Skills Test (CCTST) have become widely accepted as reliable measures of student learning. In today's learning assessment discourse, therefore, advocates are rarely called upon to defend the measurability of learning.

Instead, what one finds in LOA's audit narratives is a tendency to present audit as a universal solution to the crisis in higher education and in particular as the sole solution to limited learning and low graduation rates on American campuses. This representation is supported by the force of US federal education policy, which for the past three decades has promoted LOA, alongside other forms of accountability, as the primary means for improving the educational system. After the publication of *A Nation at Risk: The Imperative for Educational Reform* (National Commission on Excellence in Education 1983)—a major federal study that highlighted the failures in US education at all levels—legislators turned a critical eye to the nation's schools. The 1992 reauthorization of the Higher Education Act (the legislation that defines US federal law related to higher education) included for the first time a requirement that colleges and universities demonstrate commitment to student achievement. This policy change resulted in most colleges and universities adopting some form of learning assessment. Even if the results would be reported no further than in accreditation reviews, LOA programs satisfied the requirement and allowed institutions to maintain access to federal funding. In 2001, President George W. Bush signed the now-infamous No Child Left Behind (NCLB) Act into law. NCLB linked public funding for elementary and secondary schools to students' results on standardized tests. Schools that produced high scores received more funding, teachers who instructed high-scoring students often earned merit pay, whereas "failing schools," those with persistently low test scores, were shut down, often against the wishes of their communities. The policy was ended by Congress in 2015 and has been roundly condemned, but the force of LOA as an approach to improving educational quality and promoting institutional accountability survived the death of NCLB, as did both optimism and worry about its potential to reshape higher education.

A key moment in establishing audit as the universal policy solution to higher education's problems came in 2005, when Bush's secretary of education, Margaret Spellings, convened a task force to evaluate and make recommendations for improving higher education. The resulting report, *A Test of Leadership: Charting the Future of U.S. Higher Education* (commonly referred to as the Spellings Report), called for a new focus on access, cost, financial aid, learning, innovation, and accountability in higher education, to be accomplished with the help of LOA programs. Though increased accountability is only one of six recommendations it made, the Spellings commission identified LOA and other university audits as the foundation and precondition for all other improvements in higher education. "Higher education must change from a system based on reputation to one based on performance," the commission contends. "We urge the creation of a robust culture of accountability and transparency throughout higher education.

Every one of our goals, from improving access and affordability to enhancing quality and innovation, will be more easily achieved if higher education institutions embraces and implements [*sic*] serious accountability measures" (US Department of Education 2006, 20). Preceding the report's conclusive call for accountability is a litany of problems dentified in higher education. The commission's primary concerns include global competition for educational excellence, the knowledge economy's workforce needs, and declining learning outcomes, as measured by literacy surveys, graduation rates, and class-based and racial achievement gaps. As any casual observer of US education knows, these problems are complex, overdetermined in their causes, and have developed over decades, even centuries. Yet audit, accountability, and increased transparency are presented as the overarching solution to all of them.

The same move happens in *LSMS*, where LOA is represented as a universal solution—or, at the least, a universal palliative—to a complex set of problems facing the contemporary humanities. In her chapter, Laura Rosenthal explains that the problems facing literary studies stem from two causes: the poor material conditions of humanities departments (adjunctification, budget cuts, declining enrollments) and internal conflicts within the discipline itself "that have made advocating for our discipline a particular challenge" (Rosenthal 2011, 184). These conflicts are the result of literary studies' recent intellectual history, in Rosenthal's view. As theory became scholars' primary focus in the 1980s and 1990s, the discipline splintered and tended to deal with new topics—more diverse literatures, new critical perspectives—not by incorporating them but by growing through accretion. At the same time, the cultural studies approach swept across departments and opened up the range of genres and texts that literary scholars could interpret. Again, rather than increasing disciplinary cohesion, cultural studies methods brought resistance to defining what exactly it means to study literature or culture.[5] In Rosenthal's view, literary studies is a discipline that now lacks "a point," a common sense of what is at stake or why we teach students about literature and culture (193). In her version of the audit-as-universal-solution trope, Rosenthal locates a potential remedy for what ails literary studies in the conversations about values, objectives, and goals that happen around LOA programs. Her version of LOA is a "broadly collaborative project in which disciplinary goals, values, and significant contributions emerge through attention to multiple assessment projects, in which a wide range of teacher-scholars come to terms with and articulate the foundational goals of their courses and their programs, and which will lead to a more vigorous and more transparent discussion of disciplinary aims for the future" (193–94).

The range of capabilities attributed to LOA here is immense. For LOA to solve literary studies' internal conflicts, a central agency would have to coordinate a bevy of projects, presumably at multiple institutions, involving faculty across ranks reflecting on the fundamental purposes of their diverse courses and programs; then at some point they would have to compare notes and determine methods for producing reports and records to make the discussion transparent. It would be a massive undertaking, but in audit narratives these expansive claims about audit's potential do not have to be realistic to serve their rhetorical function. They offer a vision of solving the seemingly unsolvable that leads into a well-established body of literature on best practices, tested procedures, boxes to be checked, committees to be formed, and staff to be consulted, all of which reinforces the potency of audit. The move from the aspirational universal solution to the everyday practicalities of assessment and audit is a quick and easy one, and it is accompanied by a discourse that constantly reminds practitioners that the decision to do assessment rather than get bogged down in debate over alternative modes of reform marks them as reasonable, pragmatic, and responsible professionals.

3.4. THE AUDITOR AS MEDIATOR

In order for audit to work within a new domain, what Power calls "relations of accountability" must be established there (1997, 5). Relations of accountability are first and foremost new or restructured relations of power. Some individuals or groups within an organization will be designated as "auditors" with the power to develop assessments and metrics; measure performances and outcomes; and, ultimately, make an evaluative judgment about the subjects of audit, its "auditees." Power does not explain in detail how relations of accountability might be created, especially when they run counter to existing or historical power relations, but his discussions of audit performativity indicate that narrative and rhetoric, alongside changes in policy or law, play major roles in making new domains over as "environment[s] of auditable performance," complete with actors taking on the roles of auditors, auditees, and stakeholders (94).

Because of the gap between audit's real operational capacities and its outsized programmatic promises, the figure of the auditor takes on immense significance. "Auditing is a form of craft knowledge," Power acknowledges; "instead of a clear conception of output, auditing is constituted by a range of procedures backed by [the auditors'] experience and judgement" (1997, 69). Public and organizational faith in audit relies on the ethos of the people who conduct it as much as on the validity, reliability, and consistent application of its procedures. The construction of the figure of the auditor is a key task of audit narratives, because standards of reasonable practice do not precede the project of audit. Rather, as Power explains, "the *production* rather than the *presumption* of practitioner common sense is the issue" (88). In learning assessment discourse, practitioner common sense is established by representing the auditor—the assessment expert, policymaker, or administrator—as a disinterested mediator, a figure who follows reason and pursues usable knowledge through the tangled, emotional world of public and professional discourse about higher education.

In his *LSMS* chapter, literature professor and learning assessment proponent Charles M. Tung (2011) adopts the ethos of a mediator within literary studies' conflicted terrain to support his claim that assessment practice has a key role to play in reestablishing English departments' strength and resilience. At its root, Tung's argument is a good one: rather than focusing on the assessment of content knowledge or skill development, which too easily falls prey to quantitative reductionism, he advocates for literary scholar-teachers to come together around a shared method, which he calls "thick reading." By placing a hermeneutic method of reading, writing, and argumentation at the center of assessment discussions, Tung explains, scholar-teachers can "backward plan" to the creation of "curricularized classrooms," thus using assessment not as a form of audit but as a tool for reinvigorating an embattled discipline (200). Tung's claim that assessment has a role to play in reviving the humanities is echoed throughout the *LSMS* chapters (though his in-depth reflection on disciplinary methods stands out as a better case for the content of assessment than most authors' chapters provide), as is the ethos of negotiation that he develops throughout his essay.

From Tung's perspective, what threatens literary studies is "our inability to cast the multiplicity of our knowledges, skills, and methods as more than a loose bag of incompatible stuff" (2011, 212). This problem is not only a "speculative" one about the future of the field; for Tung, it is also an "institutional problem" (200). As such, his argument for assessment is contextualized by the material realities of today's English departments, which house not only literary study but also rhetoric and composition programs (i.e., freshman writing). Tung's first step in projecting an ethos of mediation and negotiation is to argue for literary studies to borrow its departmental neighbor's "openness to assessment" (200). By "insisting on the relation between low 'service' [writing classes] and the loftier goals of literary studies," rather than highlighting their separation, faculty assessors should be able to "theorize more broadly and more pragmatically" about the future of the profession, as well as to close the gap between their research agendas and teaching (200–201).

Tung's second step in situating the auditor as mediator in literary studies comes when he takes on the content of literature courses and curricula and proposes thick reading as a point of commonality among the multiplicity of approaches that characterize the field. Advocacy of thick reading and curricularized, assessable classrooms is "a pluralist, multipartisan position," Tung maintains, which allows auditor-mediators to overcome the contentious debates between adherents to symptomatic, historicist, and formalist interpretive schools (2011, 207). Learning assessment enables departments to come to such a teachable dissensus by forcing faculty to put its assumptions on the table. "Saying what goes without saying involves us in the negotiation of disagreement about the relationship between what we are doing, what we idealize in that doing, and what those idealizations blind us to," Tung explains. "This negotiation, which is exemplified in the many different shapes of thick reading, is fundamental to inquiry in the humanities and . . . may be one of the best ideals of literary studies" (204). Tung portrays auditors as participating in nested mediations and negotiations. They navigate between disciplinary splits (composition vs. literature) and intellectual disagreements (historicism vs. formalism), yet manage to capture the essence of the field in this action. Tung's vision of "discourse-based," faculty-led assessment at the departmental level is, happily, a far cry from the dystopian scene of NCLB, but his representation of faculty assessor-auditors as key players in rescuing the discipline reinforces the sense that audit is the sole solution to higher education crises and constructs auditors as the appropriate leaders of reform efforts.

Commitment to scholarly inquiry is another facet of the auditor-mediator figure in learning assessment discourse. David Mazella's *LSMS* chapter recounts lessons he learned while serving on a faculty senate committee tasked with building a Center for Teaching Excellence (CTE) in response to pressures from state lawmakers and university administrators. Mazella understands his experience as demonstrating the stubborn influence of the accountability/improvement contradiction within the learning assessment movement. He describes the work of his committee as mediating between accountability-focused administrators and skeptical faculty members to achieve reforms that would truly improve student learning while secondarily bumping the university up in the rankings and satisfying regulatory requirements. "I learned early on that it was crucial for my credibility *not* to be seen as an advocate for 'accountability,'" Mazella notes (2011, 234).

Instead, Mazella's faculty senate committee discovered that its tricky task of negotiation would be made possible by approaching and representing assessment as a scholarly project of inquiry into teaching and learning. The faculty senate, he realized, already "played a crucial and mediating role as an informal professional forum for voicing faculty concerns about teaching and instruction" (2011, 230). If those discussions could be linked, through the CTE, to assessment practices, then faculty buy-in could be achieved by tapping into the faculty's existing interest in improvement of teaching. Ultimately, Mazella argues, learning assessment is a form of "organizational learning," through which an institution becomes aware of its parts, their functioning, and places to innovate and improve. "When viewed as scholarly inquiry," Mazella explains, "assessment becomes a way of continually expanding one's own (and the institution's) knowledge about teaching and learning, both in- and outside one's own classroom" (231). Directed as it is to a faculty audience, Mazella's chapter contributes to the construction of the learning assessment auditors as fellow scholars by aligning them with faculty in a project of inquiry rather than emphasizing their surveillance or management capacity.

Tung and Mazella present the best version of learning assessment advocacy. Their advice is practical and practitioner-centered, and honors the knowledge and disciplinary expertise that faculty brings to learning-oriented projects in higher education. Yet even in bottom-up models like these, the representation of auditors as mediating, scholarly peers can be problematic. It can confuse the already complex power relations among faculty and administrator-managers, and it can obscure fundamental problems with assessment methods and data. A recent article by David Eubanks, an influential scholar of assessment and assistant vice president in the Office of Institutional Assessment and Research at Furman University, revealed that many institutions' assessment programs likely suffer from low-quality data and inaccurate inferences drawn from it. Because of the vast number of institutional, program-, and department-level assessment projects nationwide, combined with mandates to demonstrate that assessment data is always being used, Eubanks argues, typical scholarly standards for measurement quality fall by the wayside. Even if faculty-auditors approach assessment as a research project, that is, the material conditions of assessment—limited institutional and human resources, time, and support—prevent adherence to strict standards of reliability and validity. "The whole assessment process would fall apart," Eubanks warns, "if we had to test for reliability and validity and carefully model interactions before making conclusions about cause and effect" (2017, 6). It remains to be seen whether Eubanks's exposé will open further investigation into assessment's logistical and intellectual flaws, but for the moment, his article makes clear that the ethos of the auditor as mediator-scholar is likely more necessary for rhetorical purposes than an accurate representation of the ease with which faculty can become assessment experts. As Espeland and Sauder (2016) observe, sometimes "looking rational can be as potent as actually being rational. Once quantification"—or, I would add, auditor status—"is associated with the virtue of disciplined thinking and of being free of biasing emotion, its symbolic value overwrites and sometimes supersedes its technical efficacy" (24).

3.5. THE AUDITEE AS BAD ACTOR
AND RESPONSIBLE SUBJECT

Both narratives and "technologies of audit produce certain kinds of knowers, ways of knowing and not knowing under particular conditions of knowledge production and social relations," to borrow Elaine Swan's articulation (Swan 2010, 486). As just discussed, the production of the auditor figure in assessment narratives may obscure the limitations of audit-based knowing in higher education. Turning to the auditor's counterpart, the auditee, we encounter a different set of concerns, this time focusing on the social relations unique to life and employment in universities. When audit comes to the university, it encounters a unique challenge: Where potential auditees have a strong tradition of autonomy and self-regulation, how can relations of accountability be established? Audit narratives allow advocates of new accountability regimes to overcome this challenge by employing two complementary representational strategies, both of which center on the auditee. First, in public-facing rhetorical performances, audit narratives discredit self-regulating institutions and professionals by characterizing them as bad actors. They reconstitute autonomous institutions and professionals as subjects of audit by demonstrating the necessity of external supervision and reform. Second, in arguments addressed to potential auditees, audit is portrayed as promoting goals and values that align with the existing ethos of the profession or institution, thereby rhetorically incorporating it into the accountability culture as a responsible subject of audit.

Public-facing audit narratives discredit self-regulating institutions by attributing an outdated sense of entitlement and privilege to them, and they discredit professional judgment by undermining the association of expertise with reason and disinterest. The Spellings Report, for instance, represents colleges and universities as bad actors by depicting them as out of step with pressing national concerns and neglectful of their students.

The report opens with praise for US universities' track record of leading the world in educational excellence over the past half-century, but the commission quickly turns to accuse institutions of now resting on their laurels rather than innovating. The commission writes:

> American higher education has become what, in the business world, would be called a mature enterprise: increasingly risk-averse, at times self-satisfied, and unduly expensive. It is an enterprise that has yet to address the fundamental issues of how academic programs and institutions must be transformed to serve the changing educational needs of a knowledge economy. It has yet to successfully confront the impact of globalization, rapidly evolving technologies, an increasingly diverse and aging population, and an evolving marketplace characterized by new needs and new paradigms. (US Department of Education 2006, ix)

Taking up the language of Clayton Christensen's disruptive innovation theory, the commission positions universities as needing external guidance to pull them out of their complacent, "self-satisfied" languor and into the brave new world of the global knowledge economy. The repetition of "It has yet to . . ." marks higher education as lagging behind the better-informed, more worldly federal auditors who recognize that "the world is becoming tougher, more competitive, less forgiving of wasted resources and squandered opportunities" (US Department of Education 2006, viii). Even worse, the Spellings commission insinuates that institutions are derelict in their duties to students, contending that "most colleges and universities don't accept responsibility for making sure that those they admit actually succeed" (vii). Colleges' interest in students wanes once their records are added to the rankings data, the commission implies. By asserting that institutions are entitled, self-satisfied, outdated, and negligent, the Spellings Report makes a powerful statement that colleges and universities are bad actors in need of audit's oversight, one that influenced public opinion on higher education even if its recommendations were only partly adopted as federal policy.

When academic professionals, typically faculty, are represented as bad actors, they are discredited by association with negative, nonscholarly traits and behaviors. Their objections to LOA mandates are depicted as matters of emotion, taste, preference, or comfort, rather than expert judgments or intellectual positions. Among the NILOA papers, Margaret Miller's (2012) essay "From Denial to Acceptance: The Stages of Assessment," a review of twenty-five years of writing about LOA in *Change: The Magazine of Higher Learning*, stands out as a particularly sharp critique of faculty and administrator reactions to the assessment imperative. Miller's descriptive framework relies on emotional depictions of faculty and administrators by drawing an analogy between reactions to LOA and Elizabeth Kübler-Ross's seven stages of grief. She begins by recounting an assessment meeting with deans and department chairs at an unnamed Virginia university:

> Increasingly frustrated by their blank stares, I finally asked them, "What would you say to a student who asks, 'Why are you making me take a foreign language?'" The provost responded, "I'd say, 'Because I said so!'" Whenever I'm tempted to think that we've made no progress on assessment, I remind myself of that meeting. Today, the stares might be hostile, but they aren't blank—and no one tries to get by with "Because I said so." (Miller 2012, 4)

In this episode, academics appear first clueless and distracted, then bullheaded, and finally acquiescent but "hostile."

Reflecting on the causes of such emotional and apparently irrational responses to assessment, Miller later hypothesizes that "some of the resistance to assessment, particularly from faculty, is also due to fear that the results will reflect badly on us as individuals" (Miller 2012, 4). Given the insecurity with which the contingent faculty majority lives, this emotional explanation seems logical, even sympathetic to the conditions of employment in contemporary universities. But Miller's tone shifts in the next two sentences: "Added to that is our collective intellectual hubris. We are shocked—*shocked!*—that anyone should question our work" (4). These comments both mock faculty reactions to external requests for accountability (*shocked!*) and at the same time allude to popular stereotypes of university faculty as public servants who resist transparency because they fear being exposed as irresponsible to their public calling. In audit narratives like Miller's, the authority of academic professional judgment is undermined by associating faculty and administrators with emotion and self-interest, in sharp contrast to the reasonable and disinterested auditors who negotiate between constituencies to perform assessment.

When assessment proponents address auditees directly, as they do throughout *LSMS*, the rhetoric they employ is very different and serves a different purpose. In such arguments, the authors appeal to their colleagues' professional and disciplinary ethos and values systems. They frame participation in audit/assessment not as compliance with management imperatives, but as responsible professional practice and even a mode of social justice activism. Assessment advocates thus set out a path by which bad actors may be reformed into responsible auditees.

Gerald Graff and Cathy Birkenstein, well-known advocates of assessment and authors of a much-taught academic writing textbook, offer such a path in their *LSMS* chapter, "A Progressive Case for Educational Standardization: How Not to Respond to Calls for Common Standards" (Graff and Birkenstein 2011). When the Spellings Report was released, Graff and Birkenstein assert, faculty and institutions were too quick to dismiss its call for assessment as mandating standardization that would be inappropriate, if not downright destructive, when applied to such a diverse sector as American higher education. Instead, they argue, proponents of democratic, accessible education should embrace standardization as a means of making the curriculum legible to all students. To convince their skeptical colleague readers to adopt the role of responsible auditees, Graff and Birkenstein remind them that, in most circumstances, progressives agree with the imposition of standards, as when they support environmental or health and safety regulations. They then target scholar-teachers' core commitment to educating diverse students in inclusive ways by emphasizing the role of standards in "helping students negotiate that diversity" of courses and disciplinary conventions (219). Ultimately, Graff and Birkenstein contend, without standardization and assessment, only "a minority of high achievers manages to see through the curricular disconnection to detect fundamental thinking skills that underlie effective academic work" and success (219). The call to become a responsible subject of audit taps into the resistant auditee's ethical commitment to dismantling structural inequalities and democratizing higher education.

By first discrediting professional judgment and institutional self-regulation, then reincorporating them into an "environment of auditable performance," the rhetorical performances surrounding LOA promote audit as a mode of governance of academic workers and institutions. Unlike the bureaucracies that have historically structured university life, governance refrains from imposing new systems of control and behavior by fiat. Instead, governance works through capacities that already exist in individuals and organizations; it activates what motivates people and then channels that motivation to particular ends. While Taylorist management would dictate how pig iron should best be shoveled, audit sets the worker free to shovel however she pleases, as long as she shovels enough. The institution's task is to measure the shoveling and to ensure that the worker is correctly incentivized to undertake continuous improvement in the quality and efficiency of her shoveling. Audit narratives give the worker a reason to keep shoveling, to continuously improve her shoveling practice, because they represent the purpose of shoveling in ways that fit with her existing system of values and goals. Audit as a mode of governance creates compliance by keeping individuals' and organizations' attention focused on the meaningful ends, rather than the laborious (even at times exploitative) means, of assessment work.

3.6. IMPLICATIONS

Audit and assessment now permeate everyday learning and research activities in higher education, making it imperative for scholars and critics to take a hard look at how audit culture and its narratives are reshaping learning and teaching. This chapter has used the example of learning assessment discourse to draw out several of the key rhetorical and representational features of audit narratives in higher education. In LOA discourse, audit is presented as a universal solution to higher education's problems, auditors appear as scholarly mediators who can navigate the contentious terrain of disciplinary and institutional conflict, and auditees are hailed as bad actors who must recognize their responsibility to students and learning by becoming active participants in assessment culture.

Perhaps the most important of LOA's effects are the limitations it places—intentionally or not—on discourse about education, pedagogy, and university reform. It can shut down conversation about the causes of higher education's problems and constrain imagination about innovative pedagogical strategies. Michael Bennett and Jacqueline Brady (2014) rightly note that assessment contributes to the challenges facing those who would have higher education function as a democratizing, egalitarian project by obscuring the material reasons for student success or failure and by envisioning universities as places where students leave their class, race, citizenship status, and all other social determinants at the door. "Blind to structural inequalities," they write, "the conservative/neoliberal discourse of [LOA] ignores the contextual situations in which students of different classes study, obscuring the fact that better educational opportunities require fundamental social change" (38). Because the focus on learning outcomes can drown out consideration of resources, inputs, and student backgrounds from planning and policymaking processes, educational reforms fail to take into account the structural, economic, and political contexts that condition student learning.

Similarly, in a response to Gerald Graff, Kim Emery (2008) contends that product-focused assessment discourse damages the fundamental nature of education and learning by limiting our imagination about knowledge and the learning process. She writes, "intellectual inquiry leads to unexpected places. Outcomes assessment asserts the opposite." "The true key to the academic kingdom," as Emery tells it, is not the ability to know what you should learn and how well you have learned it, but instead to grasp that "the future is unknown, that research will reveal surprises, that difference offers a safeguard against narrow-mindedness, that incoherence is a condition of possibility, and that knowledge is neither finite nor fixed" (259). When students and teachers focus on predetermined, predictable outcomes, that is, we cannot advance knowledge; we can only replicate the status quo, leaving little room for the emergence of new insights and identities nor for the development of the curiosity and creativity that drive scholarship and social change.

In turn, if all new ideas must be fitted into the assessment paradigm in the end, what incentives exist for scholar-teachers to innovate or to reenliven alternative modes of pedagogy? In assessment discourse, Emery continues, "knowledge is not created in the classroom, only at best transmitted there" (Emery 2008, 257). Education becomes a process of commodity distribution and acquisition rather than an evolving project and relationship between people working together toward mutual progress and discovery. The paradigm rules out pedagogies that hinge on intense interpersonal journeys and the cocreation of knowledge, like transformative, critical, feminist, or Black radical pedagogical traditions.[6] Assessment advocates would argue that any pedagogy could be compatible with the assessment of learning outcomes, because at the end of the day a student has either learned or not. But this is exactly the problem. By predetermining what learning looks like and what should be learned, educational imagination and innovation are severely constrained.

Beyond analysis of audit narratives in learning assessment discourse, my goal in this chapter has been to experiment with the possibilities opened by applying humanities methods to questions about quantitative cultures. To conclude, I would like to suggest three starting points for continuing to develop the nascent rhetorical study of audit cultures.

First, we should ask what new objects of study, questions, and types of inquiry are made available by the addition of humanistic methods— especially those of literary and cultural studies—to scholarship on social measurement. As this study shows, there is much work to be done in the analysis and definition of the rhetoric of audit. Audits (along with metrics, indicators, and targets) are not only ideological programs coupled to sets of techniques by rhetorical meta-accounts. Audits themselves are rhetorical performances composed to address particular exigencies, purposes, and audiences within particular contexts. Approached as a rhetorical performance, audit becomes legible as a worldly process embedded in social, cultural, political, and economic structures and conflicts. The example of learning assessment suggests, in fact, that the rhetorical performances of and around audit may sometimes be more fruitful targets for analysis and critique than audit's technologies. As we learn more about the technical shortcomings of other higher education metrics, like university rankings and research bibliometrics, we should also look closely at the rhetorical and narrative constructs that prop up these epistemologically failing, yet economically successful, forms of audit in the university.

Second, we should draw on humanities scholarship about justice, subjectivity, and intersectionality as a resource for examining the ethics of representation in audit cultures. Espeland's work on commensuration provides a solid starting point for investigating the values, ideologies, prejudices, and interests that not only guide the explicit creation of categories and equivalences in metrics themselves but also color the representations of people, organizations, and processes in audit narratives and rhetoric.[7] In the learning assessment example, future work could examine the representations of higher education's stakeholders in LOA audit narratives. Students are often described as passive, compliant, and needy clients, while "the public" is typically represented as an ignorant, data-hungry populist mob. Both groups' motivations are depicted as primarily economic: students want preparation for high-wage occupations and the public wants value for money when it invests in education. How do audit rhetoric's narrow constructions of stakeholder interests influence public discourse, institutional decision-making, and our collective aspirations for the future of higher education?

Finally, the cultural and humanistic study of quantification may be able to point critics and communities toward points of leverage for regaining democratic agency in the audit society. In particular, this approach draws our attention to the possibilities for intervention and resistance that lie of the moment of becoming-auditable, when narrative representations of audit, auditors, and auditees are still in formation and are thus open to contestation on the terrains of public discourse and policymaking. All signs point to learning assessment being a permanent fixture in American colleges and universities, but its lessons about the roles of quantification, audit, and policy rhetoric in higher education can and should continue to be examined by faculty, students, and stakeholders who hope to shield our institutions, our learning, and our labor from the ever-expanding reach of damaging surveillance and management metrics.

NOTES

1. For histories, see Porter (1995) and Power (1997). For the effects of quantitative cultures on people and organizations, see, for example, Espeland and Sauder (2016)

and Shore and Wright (2015). For processes of negotiation, see, for example, Gorur (2014) and Merry (2016).

2. For exceptions, see Suspitsyna (2010) and Urciuoli (2005).

3. Espeland and Sauder (2016, 28) define a "mechanism" as "the event or process that describes the causal patterns that generate particular effects."

4. As we see in the example of LOA discourse below, narrative production certainly appears to be necessary within audit cultures, but I would argue that it is not only an "intermediate" step in the installation of audit regimes in new domains. Instead, audit discourses constantly produce and repeat stories about audit's potential, why its potential is not being met, who is responsible for these shortcomings, and how to fix them.

5. Here Rosenthal is referencing an argument made by William B. Warner and Clifford J. Siskin.

6. Writing assessment scholar Asao B. Inoue is challenging this paradigm in such works as Inoue (2015).

7. For examples of this work, see Espeland and Stevens (1998), Eubanks (2018), Merry (2016), and Noble (2018).

CHAPTER 4

The Limits of "The Limits of the Numerical"

Rare Diseases and the Seductions of Qualification

Trenholme Junghans

Quantification, we are reminded by a rich vein of recent work in the social sciences, is seductive (Merry 2016). Numbers have the capacity to simplify complexity, to render disparate phenomena comparable, to facilitate hierarchical ranking, and to travel across space and time. The simplified representations of social phenomena afforded by numbers moreover carry an aura of objectivity and scientific authority, and serve a range of interests and agendas, including demands for transparency, accountability, and evidence-based practice. This literature also reminds us of the costs of quantification: when social phenomena are represented numerically, context is stripped away, complexity disappears, and nuance and particularity are lost. Not only are the resulting representations of complex social phenomena qualitatively denuded; they also have a pernicious staying power, displacing the complex entities and phenomena they are meant to describe, and giving rise to new entities, categories, and standards against which measures and rankings will be made going forward. Such is the tendency for established regimes and infrastructures of measurement to persist—resistant to challenge and possibly hidden from view altogether—that they are frequently described in terms of path dependency and inertia. So if quantification seduces it also violates, inducing negative effects that are taken to display a marked tendency toward sedimentation and irreversibility.

My aim in this chapter is to describe a case that presents a somewhat different picture. In this case, well-established quantitative techniques for regulating and valuing pharmaceutical products are being effectively challenged, with the result that their authority is gradually eroded, and their jurisdiction curtailed. Because these developments run counter to many of the by now taken-for-granted tendencies synopsized above and discussed elsewhere in this volume, the case presents a valuable opportunity to rethink—and to qualify—some of the precepts that have come to animate and anchor much critical thinking on the dynamics and effects of quantification. In particular, the case beckons us to reconsider the characterization of quantifying regimes in terms of their propensity toward sedimentation and irreversibility, and their supposed imperviousness to critique.[1] Reversing this arc of interpretation, I describe how an initially restricted challenge to quantitative techniques for evaluating drugs to treat rare diseases has gradually expanded into a more wide-ranging critique of the uses and limits of quantification in pharmaceutical assessment. In contrast with much of the literature's relative inattention to the possibility of active intervention into established regimes of quantification, this case also highlights the role played by powerful pharmaceutical interests in challenging existing regulatory paradigms and practices, and in devising and promoting alternatives that are less quantitatively rigorous.

This case also calls for an expanded field of analysis in the critical study of quantification. As will be seen, the developments described here are taking place against the backdrop of dynamic changes in the fields of biomedicine and pharmacogenomics. These changes are rendering more fluid the entities and phenomena that pharmaceutical numbers are meant to stabilize and render intelligible, with the resulting instability providing new opportunities and openings to question the representational capacities of numbers and to highlight their artifice and their flattening, distorting tendencies. These challenges to the authority of quantification also tap into a public fascination with genomics and personalized medicine in order to cast quantitatively based modes of evidence, regulation, and assessment as outmoded and restrictively static, and to promote more qualitative and adaptive alternatives as the way of the future. In these ways the analysis presented here suggests the salience of what might be thought of as "meta" factors that potentially contribute to the (in)stability and (il)legitimacy of particular regimes of quantification: ideas about the integrity of particular categories and principles of classification as they undergird specific quantifying techniques, as well as commonsense notions bearing on the very possibility and desirability of stable classifications.

4.1. LOCATING "THE LIMITS OF THE NUMERICAL" IN PHARMACEUTICAL ASSESSMENT

At first glance, pharmaceutical products might seem an odd point of entry for exploring "The Limits of the Numerical," the research initiative out of which this volume has grown. For starters, the practice of assessing the cost and clinical effectiveness of a medicine in order to determine its value (referred to as health technology assessment, or HTA) is immensely technical and complex, based on fields of expertise that include clinical medicine, pharmacology, health economics, statistics, and regulatory science, among others.[2] What's more, the kinds of numbers employed in HTA are notably heterogeneous, and they include the elaborate statistics that underpin randomized controlled trials (RCTs) and support the derivation of probabilistic assessments of a drug's efficacy, as well as the hybrid, conglomerate measures employed to reckon a drug's benefits relative to its cost.[3] This means that quantification in the field of pharmaceutical assessment is a highly variegated and multifaceted affair, making it a very different kind of object of study than the discrete forms—such as indicators, rankings, and algorithms—that feature in many other studies of quantification.

The field of pharmaceutical assessment is also distinctive as a site for the study of the *limits* of quantification on account of the defining importance of quantification to the field. Arguably to a greater degree than is true of many other entities and phenomena to which numbers are applied, drugs are to a large extent *known through their numbers*. Under prevailing regimes of regulatory and clinical practice, a drug's "life cycle" is mediated quantitatively: from the earliest stages of its development in the laboratory, through clinical trials, market authorization, to uptake by the agencies that pay for medicines and by the clinicians who prescribe them, a drug's career in the world is mediated by evidence that is overwhelmingly quantitative in character.[4] In a sense, then, we might say that a drug's therapeutic agency is intimately bound up with its numbers. Numbers bring drugs to life, and without the right numbers (in the form of compelling quantitative evidence of efficacy and value, and an acceptable price tag, for example), a potentially beneficial drug is unlikely to have much impact on the lives it might otherwise affect.

Given the centrality and defining importance of quantification in pharmaceutical assessment, where does one begin to locate its limits? As an anthropologist working on the Limits of the Numerical project, I was especially eager to approach the topic empirically—to identify a context where quantification's authority is being actively contested and to research the means whereby its limits are asserted and enacted in practice. To this end, I chose to focus on rare diseases and the drugs meant to treat them. For reasons described more fully below, rare diseases (also known as orphan diseases) tend to be stubborn, refractory entities when it comes to the quantitative techniques and statistical operations that figure so prominently in biomedicine and in pharmaceutical regulation and valuation. On account of small patient numbers, it can be very difficult to generate robust quantitative evidence of an orphan drug's effectiveness, and small potential markets have traditionally meant that pharmaceutical manufacturers are reluctant to invest in this area. As a result, people with rare diseases have tended to feel therapeutically disenfranchised. The vast preponderance of rare conditions have no therapies whatsoever and those that do exist often come with a very high price tag.[5]

In recent years, however, rare-disease patients have joined forces with the pharmaceutical industry to push back against what for them amounts to the tyranny of the quantitative. These efforts have led to the enactment in many jurisdictions of special provisions for the assessment of orphan drugs and the introduction of incentives for investment in this area. The result has been the emergence of a space of exception, where the rule of the quantitative might be effectively curtailed and where qualitative evidence that would ordinarily be excluded as insufficiently scientific is increasingly admitted. For my purposes, this orphan space of exception is a space of opportunity for empirical study of the limits of the numerical. As I would learn in the course of four years of ethnographic research among HTA bodies, patient groups, and industry representatives, the orphan space has also opened up a zone of opportunity for actors who would mount a more wide-ranging challenge to the uses of quantitative evidence in pharmaceutical regulation and assessment.[6] As will be seen, the orphan space has become an incubator and testing ground for broader critiques of quantification's rule, and a laboratory where alternative modes of evidence, regulation, and assessment are being devised, trialed, and promoted. In sum, it is a space where the power of the quantitative is being contested, and where more qualitative alternatives are being pioneered and valorized. It is also a space that affords insight into dynamics often overlooked and underplayed in the critical study of quantification.

4.2. THE CHALLENGE OF RARITY

Orphan diseases tend to be surrounded by a penumbra of uncertainty of the sort that robust quantification is meant to dispel. Defined in Europe as affecting fewer than 1 person in 2,000,[7] these conditions are complex and heterogeneous, and often debilitating and life-threatening. Of the 6,000–7,000 orphan diseases identified to date, between 70 and 80 percent are thought to be genetic in origin and their natural histories are poorly understood. At last half of all cases involve children, one-third of whom are likely to die before the age of five (EURORDIS 2020; Global Genes n.d.). Until recently, there was very little incentive for pharmaceutical companies to invest in the development of orphan drugs on account of the small markets involved and because of the challenges involved in producing evidence of their effectiveness according to the "gold standard" of medical evidence, the RCT. This started to change in the 1980s, when a number of jurisdictions began to enact provisions meant to incentivize the production of orphan drugs. Among these provisions are the granting of market exclusivity for orphan drugs (ten years in the European Union and seven years in the United States); reductions in or exemptions from fees paid to regulatory agencies; free scientific advice in the development of experimental protocols; expedited regulatory review; and (in the United States) a tax credit of 50 percent for the cost of conducting human trials.[8] These provisions are widely recognized as having stimulated orphan drug development. Thus whereas prior to the introduction of the first orphan legislation in the United States in 1983, only thirty-eight orphan drugs had been licensed (EvaluatePharma 2014), by early 2021 that number had grown to 945 (FDA),[9] and by the early 2000s a number of other jurisdictions (including Europe, Japan, and Australia) had likewise introduced provisions intended to promote the development of rare-disease treatments (Tambuyzer 2010).

Notwithstanding the success of orphan legislation, patient access remains a problem on account of the stubborn elusiveness of statistically robust evidence of the benefits delivered by many orphan drugs. High prices combined with high levels of uncertainty concerning a drug's clinical and cost effectiveness mean that orphan drugs are likely to exceed willingness-to-pay thresholds, be they formally set (as in England, by the National Institute for Health and Care Excellence [NICE]) or informally observed (as is true in a number of other European jurisdictions). In order to combat these barriers to access, HTA bodies are increasingly recognizing the need to *qualify the ways quantification is used* for assessing orphan drugs. In many jurisdictions there is increasing allowance for the admission of evidence that would ordinarily be excluded as lacking in quantitative rigor—that is, evidence gathered by means other than the RCT. Generically referred to as real-world evidence (RWE), such alternative forms include evidence from trials lacking a control group and/or involving a very small number of patients; evidence gathered retrospectively from medical registries; observational and anecdotal reports from clinicians; and the experiential accounts of patients and carers. These softer forms of evidence are regarded by many to be of limited scientific value and as inadequate substitutes for RCT-derived evidence. The very term "real-world evidence" is regarded by some as a rhetorical dodge, meant to lend credibility to substandard forms of evidence.[10]

Ethnographic research I conducted with appraisal committees in the United Kingdom (NICE in England and the Scottish Medicines Consortium [SMC] in Scotland) suggests how tensions and contradictory dynamics might follow from greater reliance on qualitative forms of evidence and nonstandard methods of appraisal. Because individual rare diseases are often poorly understood, information gleaned from patients, carers, and expert clinicians are thought to be particularly important in appraising the potential "added value" of a therapy for a rare disease. This is all the more true when the therapy has not been subjected to an RCT. The dossier on which a drug's appraisal is based primarily comprises quantitative clinical and economic data, and routinely runs 500–600 pages in length. By contrast, evidence sourced from patients and the people who care for them is often denominated in qualitatively rich and affectively powerful narrative, which can mark it as radically different from the rest of the dossier. Despite their potential value, patient-derived inputs can thus present appraisal bodies with a dilemma: how to re-present information gleaned from patients in a way that renders it assimilable with other forms of evidence yet still retains its inherently emotive quality and force. HTA agency staff responsible for preparing a drug's dossier work hard to reach this elusive balance. Notwithstanding these efforts, committee members responsible for appraising the dossier and reaching a decision are apt to experience uncertainty concerning how to weigh patient inputs relative to more traditional forms of evidence, and may be wary of being unduly swayed by its raw emotional power.

4.3. COMMENSURATION AND THE PROBLEMS OF QUALITATIVE EXCESS, ACCOUNTABILITY, AND AUTHENTICITY

These dynamics—and the fundamental ambivalence they betray—can be usefully explicated through the analytic lens of commensuration (Espeland and Stevens 1998; Hankins and Yeh 2016). Commensuration—the comparison of entities judged in the first instance to be different—is a fundamentally generative process implicated in activities ranging from the most basic acts of cognition and classification to elaborate operations of regulation and standardization, commodification and valuation. Bound up with quintessentially modernist ideals of universal convertibility and the possibility of comparative evaluation unimpeded by fundamental difference, commensuration has come to assume a positive ideological valuation unto itself. This positive valuation is reflected in the metaphorical framing of the RCT as the gold standard of medicine, a framing which both references and reinforces the prized power of the RCT to compare things otherwise considered different and to evaluate them according to a common metric and hence in relation to one another.

Yet as explained by anthropologists Joseph Hankins and Rihan Yeh, commensuration also has "a built-in tendency to failure": "As the OED emphasizes, to be commensurable is to be reducible to a common measure, to be divisible without remainder. But remainder . . . is unavoidable. . . . Commensuration creates its own excesses, 'surfeits' in Nakassis's (2013) terms" (Hankins and Yeh 2016, 22). The reason why remainder, or surfeit, is inevitable derives from the fact that commensuration requires abstraction from the particular and selective disattention to any qualitative aspects of the entities being compared that might undermine the assertions of similarity on which commensuration is based. Yet it is always the case that these qualitative excesses can reappear, intruding as discomfiting reminders of incommensurability: "forceful assertions of commensurability," Yeh writes, "feather apart in the face of a persistent remainder which they themselves evoke: the excess value . . . a qualitative difference that seems to fly in the face of comparability" (Yeh 2016, 63). Hence commensuration is "always an achievement . . . subject to authoritative ratification, and vulnerable to possible failure," with the possibility of failure being "chronic and constitutive" rather than exceptional (Hankins and Yeh 2016, 9).

Viewed through the lens of commensuration, the goal of capturing an orphan medicine's "added value" is an effort to recuperate the excesses (or surfeits) that have been overlooked or erased through the commensurating operations of standard assessment techniques. But such recuperative gestures are necessarily in tension with the ideals of perfect commensurability without remainder and comparative evaluation unimpeded by incommensurable difference. The ideal of perfect commensurability might be served by re-presenting patient input in a way that renders it more *formally* consistent with standard forms of evidence, yet such an approach risks stripping patient input of its emotive power without fully overcoming its uncertain status as evidence. What's more, lacking clear guidance on how to incorporate nontraditional evidence in what is otherwise a clearly signposted and rule-governed process of assessment, committee members might be inclined to bypass it in favor of quantitative evidence, which can be more readily interpreted. In sum, the curtailment of the qualitative disrupts smooth and seemingly automatic commensuration, revealing that which is incommensurably qualitative. When this occurs, troublesome issues of accountability and authenticity are apt to come to the fore.

To understand why quantitative curtailment might beget a crisis of accountability, it is useful to bear in mind the symbolic importance and moral value that have come to attach to the use of quantification in both science and public administration. According to historian Lorraine Daston (1995), the moral economy of science demands "the reining in of judgment, the submission to rules, the reduction of meanings," all of which are practically and symbolically enhanced by reliance on quantification (10). And as powerfully argued by Ted Porter (1994), "among the most important roles of statistics in science . . . is to render an accounting of belief" (389; see also Porter 1992, 1995). In the nineteenth century this scientific ideal of impartiality also came to infuse an ethic of public service, such that quantification became the symbolic guarantor of technocratic virtue and trustworthiness:

> The appeal of numbers is especially compelling to bureaucratic officials who lack the mandate of a popular election, or divine right. Arbitrariness and bias are the most usual grounds upon which such officials are criticized. A decision made by the numbers (or by explicit rules of some other sort) has at least the appearance of being fair and impersonal. Scientific objectivity thus provides an answer to a moral demand for impartiality and fairness. Quantification is a way of making decisions without seeming to decide. (Porter 1995, 8)

Beyond the problem of accountability, the introduction of qualitative evidence also begets anxieties and ambivalence about the quality of that evidence and the credibility of the patient voice. The warrant of credibility for this kind of evidence is authenticity, which can itself be a difficult quality to reckon. The problem is made worse by the fact that rare-disease patients can be important strategic allies for the pharmaceutical industry, and regulators are vigilant about the possibility that vulnerable patients and their family members might be captured by pharmaceutical interests. Indeed, because these potential allies, like the diseases with which they are associated, are *rare*, pharmaceutical companies vie with one another to enroll them as partners and collaborators from the earliest stages of an orphan drug's development. Such is their authority as sources of information on the nature of the disease experience, and on what is important in terms of therapeutic endpoints, that patients and their representatives are as much valuable assets as strategic allies, and industry relates to them accordingly. While conducting participant observation at a week-long summer school to train rare-disease patients and their representatives, I was struck by warnings given about the potential risks of working too closely with the pharmaceutical industry. The annual summer school is run by EURORDIS, an independent alliance of European rare-disease patient groups, with the aim of providing participants "with the knowledge and skills needed to become experts in medicines research and development" (EURORDIS n.d.). In other words, the course helps patients and their families parlay their personal experience into recognizable expertise for purposes of advocacy. Participants were repeatedly warned, however, of the dangers of overly close collaboration with industry, as this would compromise their credibility in the eyes of regulators, and disqualify them from participating in the deliberative processes.[11]

The sensitivity around the issue of authenticity of "the patient's voice" would subsequently be affirmed for me in the context of another ethnographic encounter, this time at an international conference dedicated to orphan drug development. This annual two-day event is advertised with the tagline "strategy, advocacy, and partnering for the orphan disease industry," and the "early bird" price of admission is close to €2,000 (or $2,200), with discounts available for academics and regulators. Rare-disease patients and their representatives are eligible for free admission and are in some sense guests of honor: personifications of the nobility and urgency of the orphan cause. One conference panel comprised three North American "patient experience officers" from different pharmaceutical companies, and each spoke with exaggerated obsequiousness about rare-disease patients and their family members as the foremost experts in their own conditions, with vitally important roles to play throughout the "life cycle" of an orphan drug. Recalling the warnings I had heard given to aspiring patient experts, I ventured to catch the eye of the jovial facilitator with the roaming mic. "But what about the risk of patient experts being perceived as corrupted by industry influence, and disqualified as experts?" I asked with strategic indecorum. The mic was quickly snatched from my hand and one of the panelists tersely dismissed my query before moving on to the next question: "*That*," she angrily snapped, "is *not* the kind of language we use here."

If the rare-disease patient can be matter out of place in the regulatory and assessment processes—an awkward surfeit not readily assimilated—my skeptical intervention in this context was likewise an unwelcome excess, an excrescence threatening to defile a presumably lofty enterprise. More critical agent than docile patient, I along with my intervention provoked angry denial and symbolic expulsion, rather than the sympathetic recognition and inclusion that defines the spirit of orphan exceptionalism.

In the foregoing sections I have described the coming-into-being of a space of regulatory exception, where the rule of the quantitative is abridged to accommodate rarities that are not easily governed by established quantitative techniques. I have suggested how orphan provisions can cause hiccups and inconsistencies in a domain that prizes regularity, uniformity, and commensurability. These provisions can also feed concerns about accountability and anxieties about the authenticity of the patient voice. In the next section I will describe how orphan exceptionalism has also given rise to a parallel anxiety in the form of skepticism about the authenticity of the orphan disease *and* the authenticity of the orphan drug: When is a condition "truly" rare, and how distinctive does a therapeutic compound need to be in order for it to be deserving of orphan recognition? As will be seen, there is now something of a gold rush underway within the pharmaceutical industry to enter the "orphan space," creating a need to police its borders, interrogate claimants to orphan status, and expose pretenders. This is occurring against the backdrop of developments in pharmacogenomics that have made it possible to therapeutically target genomic subtypes of common conditions, with the result that the number of "rare" diseases is increasing. In the process, established disease categories and classificatory conventions are being upended, creating wide latitude for debate and maneuver.

Taken together, these developments have a twofold significance for the critical study of quantification. First, skirmishes and debates around *just what constitutes authentic rarity* direct critical attention to the vital importance of *classificatory schema and conventions* in the workings of quantification. Second, these developments remind us that if things must be "made quantitative" (Porter 1994), then by the same token entities that have been rendered docilely quantifiable might be requalified as less amenable, resistant, or impervious to quantifying operations.

4.4. POPULATING THE ORPHAN SPACE: ORPHANIZATION, STRATEGIC SINGULARIZATION, AND SALAMI SLICING

The commercial and regulatory benefits associated with having a drug treated as an orphan have created strong incentives for pharmaceutical companies to manufacture products that can inhabit this charmed space of exception. A 2017 newsletter from the industry consultancy EvaluatePharma provides details on the "steady and inexorable growth of the orphan drug market," noting that growth in sales for the orphan sector is forecast to be 11 percent annually, or more than double the rate predicted for conventional drugs. The median cost of orphan drugs per patient is 5.5 times higher than that of nonorphans, and by 2022 orphan drug sales are projected to total $209 billion worldwide and to comprise 21.4 percent of prescription sales. The US Food and Drug Administration (FDA) is receiving record numbers of applications for orphan status; of the fifty-three new therapeutics approved by the FDA in 2020, thirty-one (58 percent) were for rare diseases; and orphan drugs are forecast to comprise 55 percent of the cumulative European pharmaceutical pipeline by 2022 (EvaluatePharma 2017a). Pharmaceutical industry conferences often headline topics such as "Taking Maximum Advantage of Rare/Orphan Disease" and "Driving Growth by Building the Global Leader in Rare Conditions" (both at Fierce Biotech Drug Development Forum, Boston, September 2016), and oft-cited publications include "Orphan Drug Development: An Economically Viable Strategy for Biopharma R & D" (Meekings et al. 2012) and "The Economic Power of Orphan Drugs" (Thomson Reuters 2012).

In practice, the pursuit of orphan advantages entails what is sometimes referred to as *orphanization*: positioning products in such a way that they can be plausibly considered to treat a rare disease and/or can be considered sufficiently distinct from existing products to gain rights of market exclusivity. These maneuvers, which might also be thought of as instances of strategic singularization, have led to skirmishes over what ought to count as a "rare" disease for the purposes of orphan designation and uncertainty about what criteria should guide determinations of sameness and difference for purposes of classification, comparison, and valuation. There are well-known cases of pharmaceutical companies receiving approval under orphan provisions for a drug that turns out to have therapeutic applications for a much broader spectrum of conditions than the one for which it was initially authorized (Cheung et al. 2004; Loughnot 2005). In this way, drugs for rare diseases can, ironically, become "blockbusters."[12] Additionally, the pharmaceutical industry can benefit from subdividing common disease entities into smaller subtypes, which can effectively be categorized as new rare diseases. Regulators refer to such tactics of subdividing a more common disease into subgroups as "salami slicing" and often bemoan the practice as an abuse of orphan provisions (Herder 2013; Loughnot 2005).

Opportunities for salami slicing have vastly expanded with advances in pharmacogenomics, which are enabling disease categories to be subdivided into ever more granular genomic subtypes (Kesselheim et al. 2017). In the words of one legal analyst, pharmacogenomics takes salami slicing to a new level, such that "each genetic variation could possibly be considered a disease unto itself" (Loughnot 2005, 374). In recent years an average of 250 new rare diseases have been "discovered" annually (Herder 2013, 245–46), and no less an authority than the EMA's senior medical officer, Hans-Georg Eichler, has prognosticated that we are fast approaching a time when *all* diseases will be rare diseases.[13] These developments are also destabilizing established schema of disease classification, as "phenotypically homogeneous diseases are more frequently becoming 'subsetted' on the basis of genomics [and] conversely, overlap of therapeutic mechanisms of action is increasingly seen across seemingly diverse diseases" (Maher and Haffner 2006, 71).

Although these trends bode ill for the future affordability of medicines, it can be difficult for regulators and payers to oppose them without appearing to be rigid and sclerotic in the face of brave new developments in pharmacogenomics and biotechnology. There are, moreover, significant resource asymmetries between regulators and pharmaceutical companies, as the latter are well positioned to make convincing claims about what constitutes rarity and to promote classificatory regimes from which it stands to accrue benefit.

In light of these developments, some experts are concluding that orphan drug legislation may have outlived its usefulness and is at the very least in need of reform.[14] It has been suggested that orphan recognition might be granted not on the basis of the size of the patient population for a particular rare disease but rather on the basis of the total number of patients (possibly with different conditions, rare and common) treated by the same drug. Another proposed change would involve using as the basis of orphan designation the genomic target of the drug in question (Hogle 2018, 53). Still another possible modification would entail withdrawal of orphan benefits if revenues for an orphan product exceed a specified threshold. While measures such as these might serve as a brake on some abuses of orphan provisions, they do little to address more fundamental challenges that arise in conjunction with advances in pharmacogenomics and biotechnology: What should even count as a disease? And what is a drug? For example, when a drug's uniqueness is being assessed for purposes of granting rights to market exclusivity, what should be the most relevant criteria—active ingredient, for instance, or the entirety of the compound? Should diseases be classified on the basis on phenotype or genotype, or something else? And what of new kinds of pharmaceutical entities such as stem-cell therapies that straddle established binaries and boundaries, and that disrupt the organizing work they perform (Hogle 2018; Holmberg 2011)?

Crucially, the pharmaceutical industry is highly attentive to the risk of regulatory pushback and is very proactive in its efforts to gain advantage in ongoing regulatory skirmishes, and to seize and hold the high ground in this dynamically shifting terrain. EvaluatePharma's 2017 orphan drug report baldly attributes growth in the sector to "the current willingness of payers to stump up for the huge price tag" of orphan drugs, but warns of an impending backlash, as some payers are starting to insist on "much narrower interpretations" of the clinical effectiveness of orphan drugs (EvaluatePharma 2017b, 4). In its 2018 report, the same consultancy reiterates its observation that "the pushback" has begun, as both payers and politicians are scrutinizing orphan treatments more closely and "asking questions about the fairness of big pharma of using the associated tax and regulatory advantages of developing orphan drugs" (EvaluatePharma 2018, 6). In light of these threats, the 2017 report concludes, "the orphan drug industry must continue to generate innovations that justify the huge cost of these life-transforming treatments."

Taken together, these developments present a valuable opportunity to observe factors and dynamics that are insufficiently addressed in the literature on quantification: the role played by classification in rendering entities more or less amenable to quantifying operations and the possibility that interested parties might actively maneuver and intervene to influence the classificatory schema and the conventions and habits of discernment that guide practices of classification, counting, and commensuration. Because such maneuvers and interventions have the potential to fundamentally disrupt or enhance operations of quantification, they warrant critical attention. To better understand these dynamics, it will be useful to consider what is involved in "making things quantitative" (Porter 1994), as these entailments can provide a roadmap for thinking about how things might be made *less amenable* to quantification. This will in turn provide a foundation for the chapter's concluding discussion of ways in which actors in the pharmaceutical industry appear to be actively maneuvering to sway hearts and minds in ways that will serve their interest in further departures from current regimes of regulation.

4.5. [UN]BECOMING QUANTITATIVE: CLASSIFICATION, COMMENSURATION, AND THE QUANTIFYING DISPOSITION

Historian Alfred Crosby reminds us that the propensity to count and the compulsion to measure are historically conditioned. Reflecting on a Brueghel canvas, Crosby discerns evidence of the waxing of just such a quantifying bent in northern Europe in the mid-sixteenth century:

> Many of the people in Brueghel's picture are engaged in one way or another in visualizing the stuff of reality as aggregates of uniform units, as quanta: leagues, miles, degrees of angle, letters, guldens, hours, minutes, musical notes. The West was making up its mind . . . to treat

the universe in terms of quanta uniform in one or more characteristics. (Crosby 1997, 10)

To echo Daston's work in historical epistemology (1992, 1995), we might add that the painting's title—*Temperance*—underscores the ethical weighting of the quantifying impulse during the same period and its coding as morally virtuous.

Crosby also reminds us of the crucial but often overlooked fact that *measurement requires counting*, and that counting makes sense only if things are judged to be *sufficiently similar* to justify counting them: absent a sufficient degree of similarity—which is to say absent the possibility of categorizing things as belonging to the same class—counting can be something of a fool's errand, a case of the proverbial mixing of apples and oranges. It follows that if parsing the world into uniform units ceases to seem plausible, natural, and desirable, so too might quantifying operations lose some of their seductive power.

To further unpack what is involved in making things quantitative—and hence how quantification might come to seem less plausible—it is worth noting that far from being an exclusively technical operation, the parsing of the world into uniform units also depends on fundamental epistemological and ontological commitments that vary across space and time, and that cannot be taken for granted. Plato, for example, was heir to a Heraclitean sensibility of constant becoming and flowing change, and Aristotle famously discerned the artifice and violence of measurement, the fact that it can proceed only by qualitative denuding, the "[stripping] off all the sensible qualities, e.g. weight and lightness, hardness and its contrary, and also heat and cold and other sensible contrarieties" (Crosby 1997, 13). As suggested by Crosby's reference to the ancients, we as critical scholars are accustomed to recognizing ontological and epistemological commitments of the type that bear on operations of quantification, classification, and commensuration within the philosophical corpus. Yet it is also important to recognize that analogous commitments likewise obtain in what might be thought of as the everyday vernacular, in the form of habits and dispositions of judgment and discernment, and in quotidian acts of classification and commensuration. Such habits and dispositions warrant our critical attention because they play a crucial role in framing quantification as virtuous or otherwise.

To illustrate what I have in mind when I refer to vernacular analogues to philosophical commitments bearing on the possibility and propriety of quantification, consider first an explicitly philosophical articulation: "Similarity," Nelson Goodman once claimed with impish flourish, "is insidious": "Similarity, ever ready to solve philosophical problems and overcome obstacles, is a pretender, an impostor, a quack. It has, indeed, its place and its uses, but is more often found where it does not belong, professing powers it does not possess" (Goodman [1972] 1992, 13). Translated into a more prosaic idiom, Goodman's point was that ascriptions of similarity can be epistemologically and philosophically fraught because any reckoning of similarity will always be partial and consequent on selective attention to certain aspects of objects under consideration and on disattention to other aspects (something also observed by Aristotle, as noted by Crosby above). Thus Hahn and Ramscar (2001, 3) comment:

> As Nelson Goodman (1972) famously pointed out, similarity alone might be taken to be an empty explanatory construct: any two things can be similar or dissimilar as you like, depending on the respects in which their similarities or dissimilarities are described. The pigeon outside my window, the chair I'm sitting on, and the computer on my desk share numerous similarities with respect to their closeness to me, and their distance from the sun, and so on. They all share numerous dissimilarities in respect of their animacy, or lack of same, or with respect to whether they are actually in my office or not, etc. Unless we specify the *respects* in which things are said to be similar, the act of saying that they *are* similar is an empty statement.

Moving away from the domain of philosophy proper, analogous tenets and commitments have also animated debate and controversy in the histories of medicine, classification, and taxonomy, with positions ranging from the belief that classificatory systems could "carve nature at its joints," to the nominalist view that "species, genera, or families are just names used for us in grouping individuals according to our explanatory and pragmatic goals" (Huneman et al. 2015).

   It seems to me that we can discern a vernacular version of these kinds of concerns and commitments when people display greater or lesser degrees of comfort or skepticism with respect to mundane acts of classification and commensuration, and in everyday assertions of similarity and difference. No less than professional philosophers, scientists, and taxonomists, human subjects in their everyday activities will display conditioned propensities toward lumping or splitting, and differential inclinations to discern difference or similarity. We might think of these dispositions and inclinations as akin to Pierre Bourdieu's notion of *habitus* in that they are patterned and habituated; inculcated yet adaptive.[15] Yet more so than *habitus*, perhaps, predispositions of this type are malleable and liable to shift with context and according to a subject's pragmatic ends and agentive acts. Is the time you spent washing dishes really equivalent to my having shopped for groceries and prepared dinner? Is one black coat interchangeable with another or does it matter that the first one is made of cashmere and the other of merino? Are Mexican pesos exchanged for US dollars at the official rate equivalent or does it matter that one currency is likely to depreciate at a faster rate than the other?

With respect to the material presented above, the genomic subtyping of common diseases that is feeding an increase in the number of rare conditions might be regarded as part of a shift in patterned practices of classification and perception in the context of dramatic sociotechnical changes in genomics and biomedicine. Pharmaceutical efforts to position new products in the orphan space of exception by means of strategic singularization—emphasizing or materially enhancing a product's uniqueness compared to anything else with which it might be plausibly grouped and against which it might be assessed and valued—are examples of strategic, agentive interventions into habits and practices of classification in order to gain commercial advantage. When patient activists and their corporate pharmaceutical allies push back against the tyranny of the quantitative, they, like Goodman, are refusing and exposing the flattening artifice and abstracting violence of quantitatively grounded forms of evidence and paradigms of regulation. All might be seen as examples of the disruptions that might ensue, and the advantages that may accrue, when the qualitative excess that makes entities rare, unique, or even singular is recuperated.

To sum up—and to speak as one who has been enculturated to "naturally" quantify—quantification requires measurement, which requires counting, which is based on operations of classification and commensuration. These are culturally and historically variable practices and subject to contestation and change. Orphan incentives and regulatory exceptions, along with developments in pharmacogenomics and biomedicine, have provided the pharmaceutical industry with both motive and opportunity to intervene in pursuit of commercial gain: to rejigger categories and classificatory schema, and to act upon the habits of judgment and the perceptual repertoires which render quantification more or less natural and legitimate. As will be shown in the next and final section of this chapter, industry actors are intervening not just among regulatory experts and other professional peers, they are also working to influence vernacular habits and dispositions, and to shape a new cultural imaginary which is more attuned to the provisional nature of categories and classification, and more skeptical about the affordances of quantification.

Before briefly sketching how actors in the pharmaceutical industry are working to shape a cultural imaginary wherein the seductive powers of quantification are diminished, it is worth noting the significance of such developments for this chapter's larger argument. These developments indicate the possibility of dynamic and agentive intervention into a well-established regime of quantification. They also throw into sharp relief what appears to be a bias in the critical literature on quantification toward viewing regimes of quantification in terms of their presumed propensity to sedimentation and irreversibility (Bowker and Star 2000; Merry 2016; Rottenburg et al. 2015). Without discounting the distinctiveness of the pharmaceutical case—namely the involvement of very well-resourced actors playing a high-stakes game in the context of dynamic changes in biomedicine—this bias nonetheless warrants critique. Although such a critique is well beyond the scope of the present chapter, it bears noting that it seems to be partially informed by selective attention to the material dimensions and built infrastructures of quantifying regimes and a corresponding disattention to its more ideational, cultural, or dispositional aspects.[16]

4.6. QUANTIFICATION QUALIFIED: ADAPTIVE BECOMING, ANTICIPATION, AND THE NEVER SELFSAME

In this final section I describe how the orphan space is being actively cultivated from within as a zone of regulatory and dispositional experimentation. This is occurring through a spectrum of pilot initiatives that deviate from the logic of established regimes of pharmaceutical regulation and that instantiate and valorize sensibilities other than those that support quantitatively grounded modes of regulation. These budding sensibilities celebrate adaptation and emergent potential over stability, regularity, and standardization. Arising in the wake of developments in the fields of genomics, synthetic biology, stem-cell research, and tissue engineering, these sentiments and orientations are coalescing into what might be thought of as a "postgenomic" cultural imaginary. Such an ethos is a valuable cultural touchstone and resource for commercial and industrial actors who are seeking to push back against onerous norms of regulation, assessment, and valuation, and to gain advantage in defining a new regulatory landscape. Born of sociotechnical changes that are likely to be popularly perceived in terms of their emancipatory potential, identification with this emergent dispositional ethos also enhances claims to paradigmatic novelty and lends credence to regulatory norms that more staid and cautious temperaments would disparage as insufficiently rigorous.

Medical anthropologist Linda Hogle (2018) has aptly noted a growing incongruity between fluid entities of the sort that are emerging in the
fields of bioengineering and regenerative medicine, on one hand, and
the stable categories on which current regulatory regimes are based, on
the other (53). New entities such as stem-cell therapies and engineered tis-
sues have been referred to as "bio-objects" and as "boundary crawlers," in
recognition of their "unruliness" when it comes to established categories,
classificatory habits, and regulatory norms (Haddad et al. 2013; Webster
2013, 5). Because "living" biotech entities and interventions such as stem
cells and gene therapies cannot be easily regulated by the standardized
and standardizing forms of the RCT and evidence-based medicine (EBM),
they are providing an opening and a charismatic platform from which to
critique existing regulatory norms and to promote industry-friendly alternatives. In this way, and as already suggested, unsettled classificatory
and regulatory boundaries have opened spaces of opportunity. In availing
themselves of these opportunities, commercial and industry actors are
seeding the regulatory terrain with what are known as "smart" models of
regulation, based on constant adaptation and iterative experimental cy-
cles. Often promoted in the name of "disruptive innovation," and exploit-
ing popular expectations of novelty (Webster 2012, 7), such smart, adaptive forms of regulation depart sharply from the principles and precepts
that inform RCTs and EBM. To paraphrase Hogle's summary of these differences, "smart" regulatory modes prize speed, efficiency, and flexibility
over cautious control. They also endorse the continuous collection of data
in real time, favor probabilistic statistics over their causal counterpart,
and promote adaptive trial designs. Amenable to small trial sizes (possibly of the "n of 1" variety), where the patient is her own control, they also
endorse the use of evidence based on observation or computational models. Instead of striving for *certainty*, proponents of these new regulatory
modalities advocate for a "spectrum" of evidence, including evidence that
is merely "persuasive, promising, or preliminary" (Hogle 2018, 63–66).[17]

Amid these fast-moving developments, the orphan space is being re-
configured from within as a laboratory of radical experimentation, the
results of which may have ramifications well beyond the confines of this
protected enclave. Brought into being as a *space of exception* to accommodate entities whose small numbers made them inert and unresponsive
with respect to the quantitative norm, the orphan space now appears to
be serving as a sheltered domain where "smart" and "adaptive" forms of
regulation are being devised and trialed. Such initiatives might involve
granting new drugs conditional approval for limited use (i.e., for well-
defined patient populations) on the basis of "immature" data, with the

proviso that more data be collected "in real time," and with the understanding that approval will then be expanded or contracted as RWE accrues. Thus, conditional approval might be granted, for example, for a drug for a rare form of cancer in spite of a weak evidence base, and that approval extended to more types of cancer as evidence of the drug's effectiveness accrues. Conversely, if evidence gathered fails to demonstrate efficacy, the theory goes, approval could be withdrawn. Similar schemes are being developed for the financing of new and expensive therapies, whereby payment would be correlated with the proven effectiveness of a therapy once it has been observed in actual use.

Proponents insist that such arrangements will increase access for patients in desperate need, reduce the time and cost of drug development, and even enhance pharmaceutical innovation and sustainability. One such initiative, "Adapt Smart," is supported by a consortium of pharmaceutical companies and seems to promise the world. The initiative "seeks to foster access to beneficial treatments for the right patient groups at the earliest appropriate time in the product life-span in a sustainable fashion" (Adapt Smart n.d.). While ostensibly being trialed in the name of rare-disease patients and others lacking good therapeutic alternatives—that is, for use in *exceptional* cases—proponents of these "adaptive" regulatory pathways clearly view them as the way of the future and look forward to a time when they will be the "new normal."[18] Critics on the other hand vociferously denounce such schemes and view them as part of a "race to zero evidence" in pharmaceutical regulation and as a means to adapt regulation to the needs of industry (Davis et al. 2016).[19]

Despite the fact that these pilot initiatives have so far failed to produce unambiguously positive or conclusive results, there is reason to believe that support for them is on the rise, and that their animating sensibilities are gaining traction. In the United States, the Obama White House's 2015 "A Strategy for American Innovation" called for a "shift away from technologies that can be regulated in accordance with stable categories to technologies that enable and require more fluid approaches" (National Economic Council and Office of Science and Technology Policy 2015, 117). In like vein, FDA commissioner Scott Gottlieb in 2018 endorsed more adaptive "life cycle" approaches to drug development and regulation (Gottlieb 2018).

In the same way that the orphan space has become something of a safe haven for experimentation with less quantitatively rigorous modes of pharmaceutical regulation, so too does it appear to be a fertile ground for the cultivation of sensibilities that differ markedly from those associated with the moral economy of quantification. Put differently, the pilot initiatives being nurtured in the orphan space may also be reshaping the dispositional orientations and conditioned habits of perception and judgment that support the hegemony of quantification. Taking the affective pulse of these "smart" and "adaptive" modalities, Catherine Montgomery (2016) has characterized an ethos that favors adaptation over standardization in medical trial design in terms of a *moral economy of anticipation*:

> Anticipation references a mode of "*thinking and living toward the future*" [Adams et al. 2009] in which new opportunities are created to reconfigure the possible. Key to this is an *abandonment of the actual* for modes of knowledge production which claim to know the future, such as prediction, speculative forecast and computer simulation. Certainty is no longer the currency of this moral economy, but rather predictable uncertainty, which both fosters preparedness and incites perpetual striving for new ways to know the future. (5; emphasis added)

Anthropologist Karen Taussig and colleagues (2013) have noted the centrality of the concept of *potentiality* in the biosciences and in popular consciousness at the turn of the twenty-first century. They argue that the conceptual apparatus of potentiality allows subjects to talk about that which does not and may never exist, and opens an epistemic space that is full of unknowns, yet which might nonetheless be infused with an aura of promise and possibility. "In biomedical practices," they write, "potentiality indexes a gap between what is and what might, could, or even should be. Such a gap opens up an imaginative space of magic and mystery in which future-building activities related to animating bodies and extending life in new ways loom large" (S5). With growing attentiveness to the possibility that "things could be other than they are," the world is increasingly experienced as "a process of becoming" that "cannot be captured in universal formulas" (S6). It might be added that once a premium is put on emergent change and constant becoming—a kind of vernacular Heracliteanism—stable classification can become less important or at least can be viewed as only ever provisional. What is more—and to think with Crosby—if things cease to be selfsame through time, quantifying operations will also lose some of their solidity and salience. Unknowability comes to be coded as positive, as a condition of mystery and potentiality, and as a resource or asset to be strategically deployed.[20] If quantification has been valorized on account of its ability to banish uncertainty and replace it with positive knowledge, this growing attunement to the affordances of potentiality and the unknown might work to diminish quantification's authority and weaken its seductive powers.

Returning to the larger arc of this chapter's argument, developments in the orphan space of exception appear to be coalescing into a successful challenge to the authority of quantification in the domain of pharmaceutical regulation, assessment, and valuation. For this reason, the case presents an instructive counterweight to a tendency in much of the critical work on quantification to characterize quantifying regimes in terms of their purported propensity toward sedimentation and irreversibility, and their supposed imperviousness to critique. Developments in and around the orphan space of exception also demonstrate the ways in which powerful actors—in this case, well-resourced pharmaceutical interests— might actively, agentively intervene to challenge the authority of established regulatory paradigms and practices, and to promote alternatives that are less quantitatively rigorous. In doing so they draw upon the moral authority of people with rare diseases and tap into popular imaginings of a postgenomic future, where standardized, quantitatively tractable medical interventions will give way to treatments that are tailored to the singularity of an individual's genome.

The case presented here also argues for an expanded field of analysis in the critical study of quantification. Specifically, I have suggested the value of attending to the habits of judgment and perceptual repertoires that inform intuitions about the naturalness and desirability—or not—of a world parsed into quantifiable units and stable categories. Intimately bound up with the process of "making things quantitative," such factors might be thought of as part of the hegemonic apparatus of quantification. And like other hegemonic arrangements, those bearing on the power and appeal of quantification vary across space and time, and can be subject to contestation and challenge.

As a final observation, it is worth noting that this empirical case of a successful challenge to the power of quantification in the pharmaceutical domain has at least glancing affinities with the critical impulse out of which the Limits of the Numerical project grew: a skepticism about some of the work done by quantification and a commitment to moderate, curtail, or otherwise qualify its extensive uses and seductive powers. The fact that the "limits of the numerical" is being successfully mobilized as a condition of possibility by the pharmaceutical industry also suggests that overly zealous and overly generalized critiques of quantification may warrant critique and limitation in their own right.

NOTES

1. There is a distinct irony here, given that critiques bearing on the supposed tendency of regimes of quantification to become sedimented and impervious to critique themselves display a similar tendency toward sedimentation.

2. HTA refers to "a multidisciplinary process that uses explicit methods to determine the value of a health technology at different points in its lifecycle. The purpose is to inform decision-making in order to promote an equitable, efficient, and high-quality health system" (O'Rourke et al. 2020). HTA is undertaken by various bodies and agencies globally, including national and international organizations and insurers tasked with deciding which medical products will be made available to patients and/or eligible for reimbursement. In some contexts a distinction is made between health technology *assessment* and health technology *appraisal*, which can refer to different aspects of the overarching HTA processes as defined above. When a distinction is made, *assessment* usually refers to "the collation and critical review of scientific evidence" and *appraisal* to "the review of the assessment with consideration of all other (policy) factors by a committee to make a recommendation" (European Patients' Academy on Therapeutic Innovation n.d.). Here I use the more encompassing term *assessment* unless I am referring to explictly designated appraisal committees and activities.

3. The best-known such measure is the quality-adjusted life year (QALY). This is a controversial quantitative device whereby health benefit gained by a particular medical intervention is assigned a cost for purposes of assessing its value relative to other interventions. Benefits are a product of the patient's quantity of life after treatment (remaining number of years) and the quality of life they are expected to enjoy during those years. The resulting figure can then be used to compare a medication's cost-effectiveness relative to alternative treatments. See also Badano, chapter 7 in this volume.

4. Like many commodities, pharmaceutical products are often discussed in terms of their "life cycle," understood as beginning with product development, moving through market authorization, and reaching a zenith with maximum market penetration and achievement of peak revenue potential. Decline corresponds to loss of market exclusivity and the onset of competition from generics. Such anthropomorphizing figurations are rife within pharmaceutical industry discourse and seem to comprise a metaphoric transfer of attributes from the patient to the product. For example, in much the same way that healthcare professionals are wont to refer to the "patient journey," an industry consultancy's website declares, "Reaching the market is exciting and only the beginning of your compound's journey. The Life Cycle Management (LCM) team at Covance helps you realize its full potential" (Covance n.d.).

5. The median price of orphan drugs is 5.5 times greater than that of drugs for more common conditions (EvaluatePharma 2017a).

6. In technical usage, "health technology *regulation*" is distinct from HTA, and refers to the determination of a drug's safety and its licensing/market authorization, which typically occurs prior to a drug's assessment. In nonspecialist usage, however, HTA is sometimes subsumed under the descriptive umbrella of "pharmaceutical regulation."

7. In the United States, rare diseases are defined as affecting fewer than 200,000 people in the country, or about 1 in 1,500 (NIH 2021); in Japan, rare diseases are defined as affecting fewer than 50,000 people, or about 39 in 100,000 (Mizoguchi et al. 2016).

8. For a brief history and overview of orphan provisions, see Cheung et al. (2004). For a more recent overview of orphan provisions and business trends, see Premier Research (n.d.).

9. Data retrieved through the search engine at FDA n.d.

10. Such skepticism was expressed by signatories to a letter sent to the European Medicines Agency (EMA) in 2016, voicing concern about the agency's seeming endorsement of RWE in the context of a pilot program seeking to accelerate market access for new medicines (Natsis 2016).

11. The EMA, for example, actively solicits rare-disease patient representation on some of its committees and scrutinizes candidate representatives for potential conflicts of interest.

12. Sales of orphan drugs are also boosted by physicians' ability to prescribe a medicine for indications other than that for which it was approved, a phenomenon known as "off-label prescribing."

13. I have heard Dr. Eichler allude to this possibility on a few occasions at orphan drug-related conferences and workshops.

14. This is a common theme and topic of debate among regulators at conferences and other professional gatherings.

15. Bourdieu (1990, 53) describes the habitus as

systems of durable, transposable dispositions, structured structures predisposed to function as structuring structures, that is, as principles which generate and organize practices and representations that can be objectively adapted to their outcomes without presupposing a conscious aiming at ends or an express mastery of the operations necessary in order to attain them. Objectively regulated and regular without being in any way the product of obedience to rules, they can be collectively orchestrated without being the product of the organizing action of a conductor.

16. This neglect is all the more surprising given the fact that there are numerous conceptual frameworks and analytic lenses available to view these aspects and dynamics, with some issuing from scholarship which is very proximate to critical work on quantification. Work in historical epistemology, cultures of objectivity, and social histories of science all orbit around such ideational, cultural, and dispositional dimensions. Notable examples of work in this tradition include Daston and Galison (2010), Megill (1994), Porter (1995), and Shapin and Schaffer (1985). Work in the social sciences more broadly provides still more powerful concepts and frameworks for analyzing these phenomena and dynamics. In addition to Bourdieu's concept of *habitus* (1990), Williams's "structures of feeling" (1977), Foucault's *dispositif* (1980), and Gramsci's *forma mentis* (Buttigieg 1995) all provide means to grasp and analyze how social imaginaries, dispositional orientations, and habits of judgment and discernment can condition the workings and stability of systems of quantification.

17. This alliteratively phrased endorsement of what traditionalists would consider to be weak and incomplete evidence was made by Pearson (2007, 173).

18. Ethnographic observation at various orphan drug workshops and conferences.

19. The phrase is favored by Dr. Eichler. Eichler is an engaged and active supporter of Adapt Smart and is wont to use this expression as he publicly presents the case made by critics of adaptive approaches to regulation before rebutting them.

20. Linsey McGoey (2012) describes an analogous phenomenon—the strategic cultivation of unknowns, or ignorance, as a resource unto itself.

CHAPTER 5

Reading Numbers

Literature, Case Histories, and Quantitative Analysis

Laura Mandell

The relationship between numbers and interpretation has perhaps no-where been more scrutinized than in the emerging field of digital humanities (DH).[1] Traditionally, the humanities disciplines of philosophy, literature, religion, history, cultural anthropology, and the languages have constituted a richly qualitative realm. Humanists document and analyze the human experience, past and present, through texts, music, film, and art. With the advent of DH, numbers are making inroads in an area where the qualitative has been hegemonic. Digital methods are sometimes seen as invading the humanities—numbers are invading—and the pushback has been strong and stark.[2] People worry that the numerical obliterates both context and the notion of representation, and that hermeneutic methods honed over the course of centuries will be swallowed up by sociological data displayed in spreadsheets and charts.

Rather than revisiting these critiques and corresponding defenses of DH, ably delineated elsewhere,[3] this chapter examines what DH has to tell us about how numbers might work in the process of interpretation. When numbers arrive in a field with a rich qualitative tradition, they seem, by their very existence, to challenge that tradition to defend its nonnumerical methods—specifically, in the case of the humanities, to question the scientific rigor of interpretation. Philosophers in the field of hermeneutics have defended interpretation as legitimate knowledge, but in the past they did so by defining it in opposition to science. Currently, and indeed spurred by the increased use of numerical data analytics in biology, the humanities, and the social sciences, those who analyze their own field's methodologies and truth claims have begun to think about how narratives and numbers interact. For example, a workshop sponsored by the Program in History of Science at Princeton University (1999–2001) resulted in the volume *Science without Laws: Model Systems, Cases, and Exemplary Narratives* (Creager et al. 2007).

In what follows, I will examine nonnumerical and numerical interpretations of Jane Austen's *Emma* ([1816] 2003) in order to analyze how the rule-bound science of computation can possibly interact with a "science without laws." If interpretation is a lawless science, how does it work, and why is it valuable?

5.1. "INTERPRETIVE HUMAN SCIENCE," OR THINKING IN CASES

John Forrester was invited to the Princeton workshop Science without Laws because of an article he had published describing the logic employed by psychoanalysts as "thinking in cases," by which he means "case histories" (Forrester 1996). The other essays in the volume (Creager et al. 2007) similarly analyze how narratives ("models," "cases," "exemplary narratives," and "anecdotes") are used for thinking through a theory or a problem.[4] James Chandler (1998) describes how "the case" emerged historically and defines it on that basis:

> The relation of cases and casuistries to the notion of the situation has always, as far as I can tell, been a close one. A "case," most simply understood, is a represented situation, and a casuistry is a discipline for dealing with the application of principles to cases so understood. (195)

From the very beginning of psychoanalysis, however, it was not simply a matter of applying theory to cases; rather, psychoanalytic theory *emerges from* cases. And *Science without Laws* demonstrates that psychoanalysis is not alone in this—economics, history, and anthropology all extract or modify theoretical principles case by case.

In what way can a literary work be thought of as a "case" or exemplary narrative? Readers come to literature accepting its primary premise: What if? What if *x* happened? In that case, what *should* a person do? ("Should" is where casuistry comes in.) Although certainly some kind of identification is at play between readers and characters (Felski 2020), the cases of characters' situations and actions are really about *them* as distinct and different from "us." Novelistic scenes and the characters in them make up cases for thinking through life issues at a safe distance, when they are happening to someone else (who isn't real, who doesn't actually require our help), and so absent the pressing need for answers.[5] Without such pressure, with readers thinking in a play space, there is an opportunity: a work of literature can modify its readers' principles of judgment, just as psychoanalytic theory is built from and modified by cases.

Recently, the hermeneutic philosopher John Caputo (2000) has defined interpretation not against scientific facts but rather as a mode of coming to terms with facticity (42–44): when "life is hard" (55), we "have to sort our way through" a "profusion" of possible meanings (2) in order to interpret people's intentions and the significance of events. Thoughtless sorting, a kind of "fast thinking" (Kahneman 2011) is mechanical in the following way: people simply apply to the situation what they already know and think about the world. If people are thoughtful when they react, however, their current situation will cause them to examine their habitual modes of interpretation, to interrogate their validity, and perhaps even change those habitual modes based on what they have learned from a particular case in which "life is hard." Literature encourages such deliberation. Readers see what novelistic characters *are doing* in response to their situations, but the discussion (sometimes implicitly in literary criticism and often explicitly in literature classes) is about whether the characters are right to react or behave in the ways they do. As an aside, I wish to emphasize here that analyzing a literary work as the "case" of a character or characters is certainly not the only thing that literary critics do; there are many other reasons for studying literature. It is, however, relevant to the relationship between data analytics and interpretation. Recent critics involved in the cognitive science school of criticism, including Blakey Vermeule, believe that human brains are hardwired to interpret. Readers, Vermeule (2009) argues, discuss characters as if they were real people because part of what humans do as a species is try to understand the intentions of others.[6]

Vermeule discusses at length the character Miss Bates in Austen's *Emma*, noticing that she describes the gentry among whom she lives as "obliging." In fact, Miss Bates repeats forms of the word "oblige" over and over again in her dialogue: "'so very obliged,'" Vermeule says, "is her mantra" (2009, 182). Who is this woman who sees as "obliging," generous, the people to whom she is "obliged"? Miss Bates is an impecunious unmarried woman, dependent on the kindness of the society to which she is so very obliged, and a long passage early in the novel invites us to read *Emma* not only as the case of Emma, but also the case of Miss Bates. Emma's long diatribe against marriage provides a kind of blueprint for understanding how the novel unfolds. Emma tells Harriet that she would deeply regret it if she ever got married:

> "I have none of the usual inducements of women to marry. Were I to fall in love, indeed, it would be a different thing! but I never have been in love; it is not my way, or my nature; and I do not think I ever shall. And,

without love, I am sure I should be a fool to change such a situation as mine. Fortune I do not want; employment I do not want; consequence I do not want: I believe few married women are half as much mistress of their husband's house as I am of Hartfield [her father's manor]; and never, never could I expect to be so truly beloved and important; so always first and always right in any man's eyes as I am in my father's."

"But then, to be an old maid at last, like Miss Bates!"

"That is as formidable an image as you could present, Harriet; and if I thought I should ever be like Miss Bates! so silly—so satisfied—so smiling—so prosing—so undistinguishing and unfastidious—and so apt to tell every thing relative to every body about me, I would marry to-morrow. But between *us*, I am convinced there never can be any like-ness, except in being unmarried."

"But still, you will be an old maid! and that's so dreadful!"

"Never mind, Harriet, I shall not be a poor old maid; and it is poverty only which makes celibacy contemptible to a generous public! A single woman, with a very narrow income, must be a ridiculous, disagreeable old maid! the proper sport of boys and girls, but a single woman, of good fortune, is always respectable, and may be as sensible and pleasant as anybody else. And the distinction is not quite so much against the can-dour and common sense of the world as appears at first; for a very nar-row income has a tendency to contract the mind, and sour the temper. Those who can barely live, and who live perforce in a very small, and generally very inferior, society, may well be illiberal and cross. This does not apply, however, to Miss Bates; she is only too good natured and too silly to suit me; but, in general, she is very much to the taste of everybody, though single and though poor. Poverty certainly has not contracted her mind: I really believe, if she had only a shilling in the world, she would be very likely to give away sixpence of it; and nobody is afraid of her: that is a great charm." (Austen [1816] 2003, 68–69)

As a poverty-stricken spinster, and consequently poor—marriage being, at this moment in British history, the primary "business" women engaged in to secure their livelihoods—Miss Bates's obligation consists in being self-effacing and socially unchallenging (in being "very much to the taste of everybody"). In fact, we can say that she pays those to whom she is "obliged" for including her in their social world, despite her poverty, with the good-natured silliness that Emma so dislikes (Austen [1816] 2003, 69).

Reading this passage, one can see the point of the major events in the novel, which of course ends with Emma finally giving up her comfort and self-centeredness in choosing to marry Mr Knightley. Emma is as important to Knightley as she is to her father, but he, unlike her father, schools her in proper behavior. During the visit to Box Hill, Frank Churchill challenges everyone to say something clever or "three things very dull." Miss Bates is the first to respond by telling everyone that she can easily come up with three dull things to say by just talking as she does normally, and then she smiles, "looking round with the most good-humored dependence upon every body's assent" (Austen [1816] 2003, 291). Unable to resist, Emma responds by pointing out how difficult it will be for Miss Bates to limit herself to just three. When Emma refuses at Box Hill to be obliging to Miss Bates, she commits a breach of duty for which Knightley reprimands her. Emma, he says,

> "I cannot see you acting wrong, without a remonstrance. How could you be so unfeeling to Miss Bates? How could you be so insolent in your wit to a woman of her character, age, and situation? . . . Were she a woman of fortune, I would leave every harmless absurdity to take its chance. I would not quarrel with you for any liberties of manner. Were she your equal in situation—but, Emma, consider how far this is from being the case. She is poor; she has sunk from the comforts she was born to; and, if she live to old age, must probably sink more. Her situation should secure your compassion. It was badly done, indeed!" (Austen [1816] 2003, 294–95)

Emma is at fault because, fully conscious of Miss Bates's "case," she has rejected Miss Bates's form of payment for inclusion in their circle despite her celibate poverty; Emma has violated the code of noblesse oblige.

So much for the plot, and now to Vermeule's interpretation of the case of Miss Bates, which demonstrates in my view how literary cases are used for thinking. In the late eighteenth-, early nineteenth-century world that she inhabits, Miss Bates, Vermeule says, has been "assigned a [female] body and a place [her status] in a scheme in which her talents and abilities are deeply discounted." She has been dealt "a bad hand." "Why," Vermeule asks, endowing the character of Miss Bates with a full subjectivity, is she not "bitter" (2009, 181)? "Let me fancifully suggest," she continues, "that underneath all that babble, Miss Bates is silently, secretly suffering" (182). According to Vermeule, Miss Bates's mantra serves less as an indication of gratitude than as "an aggressive screen, which blocks as much as it shows" (183). Here we might question the extent to which Vermeule is *interpreting* Emma or *projecting her own* feelings onto the novel.

Before illustrating the value of the case of Miss Bates as a thinking tool, I want to stop for a moment and examine where we are in terms of thinking about interpretation as a science. It is precisely because of the possibility of "projection" that hermeneutics has been discredited as nonscientific. According to the two most compelling scientific methods, "scientific inductivism" and "falsification" (Underwood 2019, 160, citing Cleland 2001, 987), the knowledge produced by interpretation is "dubious" because it exerts no control over bias, and controlling for bias, rooting it out of results, is precisely the point of scientific method. In contrast, the point of hermeneutics, as articulated by Hans-Georg Gadamer and revitalized recently by John Caputo, is precisely to mobilize projections. It is not a *problem* that Vermeule is projecting; that is the point of the hermeneutic method.

If the case of Miss Bates were articulated merely as a fable, merely in bare plot outlines, the discussion of our conflicting interpretations would focus on us expressing our beliefs. But seeking to find evidence to prove what Austen thinks and is trying to convey allows us to explore the pros and cons of various ways of understanding the world, all the while trying to understand Austen's point of view. It is a way of thinking about habitual thinking and being without confronting it as such. Projection begins the process: "I think Austen is trying to say [reader's projection, perhaps mechanical, 'fast']." But such statements do not end the interpretation. Our complementary or contradictory projections are mobilized in the activity of interpretation, but we still have a whole novel to discuss. Reading the whole novel, its exact words syntactically connected in precisely the way that they are, involves finding out which projected interpretation it supports.[7] If a literary work is transformative, however, it will not fully support any single projection of its meaning, and the possibility of multiple "correct" interpretations opens up deliberative space. Literature invites projections for the sake of overcoming the bias inevitably informing "one correct interpretation."

5.2. WORDS, IRONY

Is Vermeule "right" about Miss Bates? Is that character deeply resentful? The evidence we have to legitimate or falsify the claim is the words on the page and what we know about the author's worldview, as derived from her novels. First, Vermeule's argument is more convincing than is visible in the abbreviated version of it presented here. The internal evidence for its veracity, all marshaled by Vermeule, is Miss Bates's incessant repetition of forms of "obliging." Miss Bates could have occasionally used other words to say the same thing. She could say that she is "grateful" instead of obliged or, instead of calling other characters "obliging," she could call them "generous." Vermeule's is a good, solid reading. But there is also evidence to contradict Vermeule's claim that Miss Bates is angry. Knightley discounts anger after Emma protests that Miss Bates did not understand her cutting remark:

> "I assure you she did. She felt your full meaning. She has talked of it since. I wish you could have heard how she talked of it, with what candour and generosity. I wish you could have heard her honouring your forbearance, in being able to pay her such attentions [i.e., to invite her to parties, teas, excursions, etc.], as she was for ever receiving from yourself and your father, when her society must be so irksome." (Austen [1816] 2003, 295)

According to Knightley, then, Miss Bates is not resentful at all; but "pay" in the passage above is a key word. Emma and her father are "able to pay" attention to the impoverished Miss Bates. Austen is writing at a time when the nobility is transforming in ways that her novels repeatedly protest: the nobility and gentry had in the past derived their prestige from honor, but that code is rapidly ceding to prestige derived from wealth (McKeon 1987, 131–33). In broad outlines, one goal of Austen's novels is to reform the gentry by calling them back to self-definition in terms of their adherence to high moral principles rather than via financial assets. In the world that Austen writes to create, everyone would gladly, willingly, and nobly treat Miss Bates kindly, without being obliged to do so, and she would have no reason to be resentful. But this is a world that Austen projects, not one she lives in and realistically represents in her novels.

It should be said here, as important evidence against Vermeule's reading, that Austen is not particularly known for generating portraits of aggressive women characters. Even suffering heroines such as Fanny Price in *Mansfield Park* and Anne Elliot in *Persuasion* do not seethe with resentment, as Miss Bates might be doing, if her repetition of the word "obliging" is deeply ironic. Austen is of course known for her irony, which drips from the infamous first line of *Pride and Prejudice*: "It is a truth universally acknowledged, that a single man in possession of a good fortune, must be in want of a wife." This sentence encapsulates Austen's goal in writing her novels, as described above: it mocks the lesser gentry living in Elizabeth Bennet's world for fortune-hunting, a mockery played out to the extreme in the character of Mr. Collins. But irony is not satire, nor parody (though the character of Collins borders on parody): Austen's first sentence is not angry. She is hovering above the neighbors of Netherfield Park, gently mocking them but also cajoling them into behaving otherwise by exposing to ridicule their matchmaking aspirations.

To argue for Vermeule's reading, however: while irony may not carry the aggression of satire, once deployed, it opens up uncertainty that cannot be contained. Irony does not present us with either one interpretation or another, either sincerity or aggression behind that "obliged": irony presents us with both interpretations at the same time, making it impossible to negate one with the other. Hence the uncontainable uncertainty introduced by irony and, I should add, its impenetrability to being analyzed objectively.[8] I now wish to explore numerical readings of *Emma*—seemingly more objective—in order to examine their value and limits.

5.3. NUMBERS, WORDS

An article in the *Smithsonian Magazine*'s Smart News (Nuwer 2012) reported on two sets of numerical findings of cultural data analytics. One is a stylometric analysis of literary style that uncovered gender differences, as explored in Matthew Jockers's 2013 manifesto *Macroanalysis*. (This was the book that, along with the work of his mentor Franco Moretti and the Stanford Literary Lab, brought text mining to the attention of literary critics.) The second comes from a blog post by Jockers that concerns *topic modeling*. Typically, stylometry examines an author's use of prepositions, conjunctions, the little words that are most indicative of a writer's style (Flanders 2005, 57), not topics, about which authors make conscious choices and so could choose otherwise. In other words, Jockers may not be justified in using Latent Dirichlet Allocation (LDA, a specific kind of topic modeling) in order to determine whether writing styles are related to physical embodiment. Nonetheless, Nuwer's article concludes that "female authorship is detectable by objective measures rather than just human intuition" (Nuwer 2012). Leaving aside the stylistic analysis, which I examine in detail elsewhere (Mandell 2019), Nuwer's title focuses on "male" vs. "female" topics: "Data Mining the Classics Clusters Women Authors Together, Puts Melville out on a Raft." Topic modeling isolated Melville because of his "topic," the first five words of which are "whale, ship, men, sea, whales" (Jockers 2011); because writing about whales is not typical of "male authors," Melville is "out on a raft," isolated from other novelists. According to this topic-modeling study, it is objectively true that Jane Austen and other women write primarily about marriage. The LDA engine identifies the following list of words as occurring together most regularly in texts by women authors:

> marriage happiness daughter union fortune heart wife consent affection wishes life attachment lover family promise choice proposal hopes

duty alliance affections feelings engagement conduct sacrifice passion parents bride misery reason fate letter mind resolution rank suit event object time wealth ceremony opposition age refusal determination proposals

We have seen how Austen's *Emma* focuses on marriage: because upper- and middle-class women living during the eighteenth and nineteenth centuries had to make marriage their careers, one could say, women novelists writing about women pay attention to it.

But while Jockers's topic modeling of novels is objectively true, it doesn't tell us very much. If the objective truth discovered by topic modeling—the one trumpeted in the Smithsonian article's title—is that Melville wrote about whales more than anyone else, why go to the trouble of amassing a dataset of novels, cleaning them for analysis (all of which could take up to a year), and running algorithms on them, which requires years to learn how to do?

An article published last year in the *Chronicle of Higher Education* echoes my point about the obviousness and cognitive poverty of "objective" truths derived by the Smithsonian from Jockers's work, and conveyed to us in sound bites. In "The Digital-Humanities Bust," Timothy Brennan (2017) asks, in the subtitle, "What Has the [Digital Humanities] Field Accomplished?" and answers "Not Much":

> [Richard Jean] So, an assistant professor of English at the University of Chicago, wrote an elaborate article in *Critical Inquiry* with Hoyt Long (also of Chicago) on the uses of machine learning and "literary pattern recognition" in the study of modernist haiku poetry. Here they actually do specify what they instructed programmers to look for, and what computers actually counted. But the explanation introduces new problems that somehow escape the authors. By their own admission, some of their interpretations derive from what they knew "in advance"; hence the findings do not need the data and, as a result, are somewhat pointless. After 30 pages of highly technical discussion, the payoff is to tell us that haikus have formal features different from other short poems. We already knew that.

Text miners do not escape "bias," Brennan argues, and the payoff of their work isn't worth the effort. To give Brennan his due, text-mining results are always to a degree what we "already knew." As Nan Z. Da (2019a, 601) puts it in the most recent assault on cultural analytics: "In a nutshell the problem with computational literary analysis as it stands is that what is robust is obvious (in the empirical sense) and what is not obvious is not robust, a situation not easily overcome given the nature of literary data and the nature of statistical inquiry." It is a paradox intrinsic to quantitative cultural analysis that an algorithm must return results coinciding with what is obvious to literary experts in order to prove that it works; if it does not, if the algorithm generates something surprising, then it is "not robust," in Da's terms.

Brennan's mistake in his critique of cultural analytics—and maybe Da's as well (2019a, 602, 604)—is the same as the Smithsonian Smart News bites: the assumption that what literary text miners are doing is searching for objective truths to supplant interpretive bias.[9] According to Brennan, Ted Underwood's text-mining work similarly has accomplished nothing, but Brennan is wrong about that. In addition to publishing traditional and digital articles and books, Underwood also keeps a cutting-edge blog that makes major inroads into theorizing digital humanities, all the while accomplishing infrastructure work that benefits other researchers: he has extracted fiction from the HathiTrust dataset (all of Google Books plus about 500,000 more items held in libraries) and created algorithms for correcting texts that have been mechanically keyed from page images, to name a few. In a recent talk at the National Humanities Center where he was a fellow, Underwood (2018) presented his work as using numbers to discover what we can learn about the history of literature, stating explicitly that his goal is not to be objective, but to use numbers to tease out perspectival differences. Underwood has used text mining to "operationalize"—that is, define and put into practice algorithmically—a definition of genre that allows one to select fiction out of the HathiTrust's roughly twenty-two million volumes.[10] People who take his work to be telling us what in the HathiTrust corpus is fiction and what is not, objectively speaking, have attacked him by saying, "you are using a faulty definition of genre." His response is, yes, exactly—the great thing about using an algorithm is that the algorithm can be changed, based on a different definition, and can generate results quickly. We can operationalize and compare the results of *n* number of definitions of genre, trying out what difference the details of genre definitions make in sorting cultural materials. Underwood also agrees with the critiques of quantocracy's alleged neutrality. Machine learning, he says, "absorbs biases." That it does so is actually a boon for literary history: "If you an historian, it's the biases of a specific time that we want to see" (see also Underwood 2019).

An early foray into "cultural analytics" that Brennan does not mention was published by Sculley and Pasanek in *Literary and Linguistic Computing*: "Meaning and Mining: The Impact of Implicit Assumptions in Data

Mining for the Humanities" (2008). Like Hoyt Long and Richard So (2016), whose *Critical Inquiry* essay is attacked by Brennan, their conclusion is not striking. After testing out on Whig and Tory texts of the eighteenth century sociolinguist George Lakoff's theories from *Moral Politics* (Lakoff 1996) about liberal versus conservative uses of metaphor, they conclude that "our experiments do not position us to make any final pronouncement concerning Lakoff's theories" (Sculley and Pasanek 2008, 420). However, "nothing" has been accomplished only if "something" is a scientific proof of a fact.[11] In fact, Sculley and Pasanek at first get results that confirm Lakoff's hypothesis and spend most of their article describing the methods they used for attempting to determine whether other factors not imagined by Lakoff could account for the results; they are trying to determine whether they got a false positive. And indeed, as they test for other possible factors distinguishing Tory from Whig discourse, those too give positive results. The summary is uninteresting because the details of what they tested *are* the accomplishment of this work, not the quotable conclusions. The "nothing" accomplished is a proliferation of possible causes for uses of specific metaphors during specific times, all of which matters not in general but in detail.

We already know that Austen discusses marriage in her novels; it is the details of what, how, and why that matter. I ran the novel *Emma* through Voyant Tools (http://www.voyant-tools.org), easy to use by anyone who wishes to experiment with and learn more about text mining (Rockwell and Sinclair 2016). In the resulting topics from the first tool, "obliged" and "obliging" cluster together in a topic that also contains "honour," "ashamed," and "regret," social feelings that would be specific to the status society about which Austen writes; the topic itself seems to center on social activities and interactions (adverbs, for example, that would be used in conversation).[12] The word "obliged" figures prominently in the whole corpus of Austen's novels, occurring 310 times, only 76 of which appear in *Emma*; "oblig*"—the asterisk-wildcard allowing us to search for "oblige," "obliged," and "obligation" simultaneously—appears most often in *Love and Freindship* (1790), with *Emma* (1815) coming in second and *Mansfield Park* (1814) third (https://voyant-tools.org/?corpus=austen&query=oblig*&view=Trends), suggesting that Austen's work as a whole interrogates the obligations associated with family and status—perhaps already obvious to the average Austen aficionado: Vermeule's chapter analyzes how and why. Voyant Tools can lead to a problem worth investigating, but the literary critic must perform the investigation.

The appearance of "obliged" in the topic modeling and counting words in *Emma* demonstrates that the word is repeated often—like Melville's mention of whales. Authors use repetition as a literary device to call attention to a theme, but that certainly does not mean that theme is at work in every use of the word. Voyant provides a Keyword in Context tool (KWIC) for distinguishing thematically significant from insignificant uses. The peruser of KWIC results for "obliged" (https://voyant-tools.org /?corpus=austen&query=obliged&view=Contexts) might be tempted to count someone saying "much obliged" as insignificant, unless they become aware that it is most often repeated by Miss Bates.

Unreflective repetition is mechanical, and mechanical uses of language such as repetition both call attention to and enact a problem. Really thinking about what one is saying involves multiple and varied formulations of it that lead to further exploration, not being stuck in the realm of one or two words. Something is impeding thought. Mechanical, unthinking repetition is a symptom, an unconscious attempt to call oneself to consciousness about something. "Oblige" leaves Miss Bates stuck in a language game that is both about generosity and maddening inequality. But can't Miss Bates just mean one sense of the word "oblige," only the positive sense? Does she have to mean them all when she uses the word?

Do the words we use necessarily mean more than we consciously wish them to mean? Stanley Cavell published a volume of collected essays called *Must We Mean What We Say?* (1969b). I submit that the collective answer of these essays is, "yes, unfortunately. We would often rather not" (see also Cavell 1979, 383). For Cavell, however, neither repetition in writing nor an author's "computation" of words need be mechanical. Cavell defines Thoreau's "computation" in *The Senses of Walden* (Cavell 1981): the text presents us with words over and over again in different contexts, drawing out the full implications of each word's meaning, in order to see what agreements about the meaning of the world, what shared assumptions, are embedded in that word. Thoreau's goal, Cavell believes, is for readers to see all the meanings of the word "practical," for instance, and then to be able to think to themselves, "but I don't mean it that way, only this way." What follows would ideally be uncertainty: "or do I?" If not, there are other words that can be used; if so, that's a discovery. The reader starts to explore all the possible meanings of, for example, the demand to "be practical"—not just all the definitions of the word, but meanings arising from usage in various contexts. Understanding those meanings allows speakers to, perhaps, decide to be impractical in specific contexts:

> The endless computation of the words of *Walden* [and *The Senses of Walden*] are part of its rescue of language, its return of it to us, its effort to free us of language and of one another, to discover the autonomy

of each [so that instead of uttering] an attempted choice of meaning, [we can express] an autonomous choice of words. . . . Immediately this means that we recognize that we have a choice over our words, but not their meaning. Their meaning is in their language; and our possession of the language is the way we live in it, what we ask of it. ("To imagine a language is to imagine a form of life.") [We recognize t]hat our meaning a word is our return to it and its return to us—our occurring to one another. (1981, 63–64)

Literature at the most basic level tries to wake us up to the form of life inhering in language, to make us aware (again) of the meaning and implications of the words we use. It uses mechanical means (repetition) to counteract mechanical speaking and thinking, what *The Senses of Walden* calls "sleepwalking" (Cavell 1981, 34, 56).

Austen intentionally repeats "obliged" and "obliging" throughout *Emma*, in the discourse of Miss Bates. Is it meant ironically by Miss Bates? By Austen? Numbers cannot tell us anything about the meaning of repetition, and they certainly cannot discern irony. Mary Wollstonecraft's novel *Mary* ([1788] 1976) would be among the novels written by women for which the word "marriage" is repeated enough to be counted by digital methods, and it ends this way, its last lines offering up the final thoughts of the dying heroine: "Her delicate state of health did not promise long life. In moments of solitary sadness, a gleam of joy would dart across her mind—She thought she was hastening to that world *where there is neither marrying*, nor giving in marriage" (68). To the eyes of the machine, Wollstonecraft is writing on the feminine topic of marriage, but this is a far cry from a Harlequin Romance or a novel by Danielle Steele. Yet if we work with computational methods carefully enough, they can productively mobilize "critical intuition."

Table 5.1 reveals the results of running *Emma* through Voyant's ScatterPlot tool, employing Principal Component Analysis (PCA) and Correspondence Analysis (CA) (https://voyant-tools.org/docs/#!/guide/scatterplot). PCA and CA cluster words together on a Cartesian plane, proximity indicating that the words are used in similar ways. The words "visit," "people," "family," "home," "friends," "came," "left," "tell," "look," and "hear" suggest social activities and interactions, and many seem like words people would use in conversation, though of course they can be used in internal monologue or indirect discourse as well: "doubt," "know," "feel," "suppose," "want," "certainly," "immediately," "quite," and "hardly." But while all of these words that operate in a similar way in the novel seem rather generic (one can imagine most of them occurring frequently in any novel), "obliged"

TABLE 5.1. *Emma* in Voyant-Tools.org

Tool	Parameters	Words clustering around "obliged" indicating similar usage
Principal Component Analysis* *I have listed words in the square from (*x*, *y*): (−3.4, −1) to (0, 5); "obliged" at (−2.4, 2)	Stopwords and names removed; Raw frequencies; 1 cluster; 3 dimensions; 110 terms	(Reading the plot left to right, top to bottom): visit people tell doubt feel look obliged family want came certainly
Correspondence Analysis* *I have listed words in the square from (*x*, *y*): (−0.08, −0.02) to (0.06, 0.25); "obliged" at (−2.5, 2)	Stopwords and names removed; Raw frequencies; 1 cluster; 0 dimensions; 120 terms	(Reading the plot left to right, top to bottom): room came immediately quite like day obliged look hear know home wish best hardly suppose deal happy friends left way

stands out: it does not strike me as a term used as frequently in most late eighteenth- and early nineteenth-century novels, while "quite" probably is. These scatterplots can clue us into unusual words that are used as if they were ordinary: Miss Bates comes, visits, looks, hears, leaves, and "is obliged"; these are all mechanical actions for her. She utters "so very obliged" in place of "thank you"—"thank" would be the ordinary word we might expect in this list. She wishes to use "obliged" to express gratitude, but "obliged" also means "required" to do something, and "unwillingly" is often implied. She cannot escape that meaning any more than she can escape the fact that generosity offered by all the obliging gentry and nobility around her obliges her to be "very much to the taste of everybody." Returning to "obliged" and "obliging" over and over again in her discourse, she is like a moth returning to the burning light. Repetition is a symptom, indicating both the desire and the inability to escape: she cannot escape *all* the word's meanings, and neither can readers.

Repetitions counted by machines have to be "read." A reading of the numerical results can be justified only through further numerical analysis (tweaking the datasets and parameters and algorithms, as I did to generate these results) and deeper reading—both of which involve interpretation. Case reasoning does not offer an alternative to statistical analysis so much as a necessary addition to it.[13]

Why necessary? Social scientists who use quantitative methods know very well that parsing the world into measurable categories which they do in designing the research model that precedes numerical analysis in part determines the answer to any query; organized otherwise, the world would appear differently (Drucker 2014, 128; see also D'Ignazio and Klein 2020). Similarly, the labor of cultural analytics includes adjusting both datasets and algorithms until they return results that are recognizable as adequate ("robust," to Da). As a practitioner and critic, Alan Liu (2013) argues that interactions between expert and machine are not a problem, but rather the necessary conditions of operation:

> any quest for stable method in understanding how knowledge is gener-
> ated by human beings using machines founders on the initial fallacy that
> there are immaculately separate human and machinic orders, each with
> an ontological, epistemological, and pragmatic purity that allows it to
> be brought into a knowable methodological relation with the other. . . .
> [D]igital humanities method—converging with, but also sometimes di-
> verging from, scientific method—consists in repeatedly coadjusting hu-
> man concepts and machine technologies until (as in Pickering's thesis
> about "the mangle of practice") the two stabilize each other in tempo-
> rary postures of truth that neither by itself could sustain. (416)

Meaning is not findable—you cannot "see" it as a find, in other words—until myriad adjustments have been made: it is human through and through. For example, in discussions of topic modeling during the six-year duration of Andrew Piper's "Text-Mining the Novel" project (https://txtlab.org/cat-egory/textminingthenovel/), participants expressed differing views about whether names and stopwords should be removed from a text before the LDA engine is applied. I tried running *Emma* through Voyant's Topic Modeling tool with and without characters' names in the text; I tried leaner and more detailed lists of stopwords and different methods for removing them, and finally decided to remove characters and stopwords using a particular method because I could see something valuable in the results: all those activities are acts of interpretation.[14] Crucially, the human/machine inter-action described by Liu (and enacted here) does not invalidate numerical results—far from it—it is the condition of their possibility.

Ben Merriman (2015) warns digital humanists against investing so much explanatory power in statistical significance: many statisticians have attacked it as insufficient for supporting an argument.[15] When "reading" results, interpretation-based disciplines have expertise in understanding human bias that goes beyond what statistical rigor can supply.

Reading numbers as a statistician and reading words as a literary critic have this in common: both attempt to address subjective projections, albeit in different ways. The goal of statistical analysis is to eliminate bias as much as possible, whereas the goal of hermeneutics is to harness it: human interpretation is the process of restoring aliveness to words by projecting meaningfulness into the world and then examining the process of projection itself in order to limit individual bias.

5.4. PROJECTION: THE NECESSITY OF THE LITERARY CLASSROOM

Discussions in literature classrooms mobilize an expertise that everyone, even the most novice reader, believes that they have: the ability to understand the psychology of human beings. Although of course they operate very differently, the goal in both psychoanalysis and discussions of literature is to transform the expertise of pat, prefabricated answers (prejudices, presumptions, forethought, biases, fast thinking, projections) back into what generated that expert knowledge in the first place: interpretive talent, an openness to questions, the capacity to continuously revise one's own habitual mode of projecting.

The theory of hermeneutics is still grounded in "what Hans-Georg Gadamer says it is, and beautifully, too, viz., a way of putting one's own horizon or standpoint 'into play' (*ins Spiel*) and thereby putting it 'at risk' (*auf Spiel*)" (Caputo 2000, 41). Gadamer (1988) insists that "the actual critical task of hermeneutics [is] separating true from false prejudices" by "mak[ing] conscious the prejudices guiding understanding so that . . . a different opinion stands out and makes itself seen" (77). With the reification of expert knowledge, another's opinion is simply wrong. The key is to mobilize interpretive talent in a venue where one's own certainties are not at risk, allowing a reader to explore another point of view. Both Gadamer and Caputo see the hermeneutic process as the task of knowing other minds, in life as in literature: "For Gadamer," Caputo says, "hermeneutics means a way to hear and welcome the coming of the other, both in person and living dialogue, and in the great texts and works of art of our tradition" (42).

Lacking in both Caputo's and Gadamer's accounts of the interpretive process in literature and in life is an insight (arguably, the fundamental insight) of psychoanalysis, brought to bear on other-mind skepticism by Cavell in *The Claim of Reason*. For Gadamer and Caputo, the problem is the otherness of other people, of not being able to know them: hermeneutics is the human activity mobilized to overcome that fundamental human finitude of being caught inside one's own mind. *The Claim of Reason* argues that other-mind skepticism is a defense against something far more frightening to people and far more commonly experienced: that sometimes others—not just friends but strangers—know us better than we know ourselves (Cavell 1979, 351–53, 383). John Forrester's example of this problem in *Dispatches from the Freud Wars* (1997) is a man yelling at his loved ones, "I AM NOT ANGRY!" (229). Of course he is, it's obvious. The man doesn't want to know that he is angry, probably from a belief that he shouldn't be, but there it is, out in the open, seen by everyone around him and, at that moment, by him himself as well. Interacting with others exposes one's own hugely embarrassing mechanicities and forces one to think them.

Psychoanalysis depends, of course, on transferring to the analyst a set of character traits and habitual, mechanical reactions that originally belonged to someone important in one's life. Its goal, and I would argue the goal of literature, is what Cavell characterizes as the goal of *Walden* and ordinary-language philosophy: to bring our words back "to a context in which they are alive" (Cavell 1981, 92; see also Ogden 1997, chap. 1). Both psychoanalysis and literature create that context by generating the curiosity, and the safety, necessary to explore the full implications of any word's meaning, no matter who said it: "The art of fiction is to teach us distance—that the sources of what is said, the character of whoever says it, is for us to discover" (Cavell 1981, 64), even when "whoever" is oneself. The analysand discovers facets of his or her own character that are beleaguered by contradictory imperatives, usually expressed as the symptoms for which he or she sought help. Ideally, the psychoanalytic process transforms the resolution of those contradictions from symptoms into art.[16]

The psychoanalytic scene is a co-created fiction. In interacting with the psychoanalyst, attributes of that important person in one's past are transferred onto the analyst. When lying on the couch, a person speaks to no one because he or she cannot see the analyst's face and reactions. There are no visual clues about what the listener is thinking, so that the analysand has to make up the reaction that he believes he is eliciting, based upon his own habitual modes of reaction. A lengthy recounting might be met by silence. "You probably think that I shouldn't be angry," he says, projecting. "Why would you think that?" the analyst asks. The analyst really wants to know what specific principles are being mobilized for interpreting intentions, one's own as well as the intentions of others. The point of psychoanalysis is "to enable the patient to know what he already knows; to re-find a talent" (Phillips 1994). In the play space of psychoanalysis, the analysand is able to mobilize his own talent for interpretation, in the process thinking through how his own, habitual, mechanical interpretations are made, and on what bases.

To summarize, then, psychoanalytic method starts with the premise that "every person investigated by psychoanalysis view[s] herself as an expert" on interpreting the psychology of others, but her spontaneous interpretations are a matter of projection, otherwise known as "cognitive bias." According to Forrester (1997), the ordinary person's "expertise" in interpreting the psychology of others is "the essential starting-point [and] the central problem" (242). Knee-jerk certainty based on commonsense modes of interpretation, as well as on principles derived from any person's particular upbringing, needs to be transformed into interpretive talent—"I know what's going on" into "How do I know what's going on, and am I right about that?" Two crucial features of this transformation are (1) nothing is learned without the knee-jerk interpretive process coming into view; and (2) "Am I right about that?" is not a question that needs to be answered but a heuristic device for figuring out how unconscious, fast modes of thinking actually work, what beliefs they entail, giving the interpreter the opportunity to decide whether he or she actually agrees with those unconscious beliefs.

Vermeule points out that novels "unmask" "cognitive biases" (2009, 177), but they do so first by creating the opportunity for mobilizing them. Everyone participating in the discussion brings forth their projections of meaning in order to discuss cases, like that of Miss Bates. "She is not angry," one student will say; yes, she is, another argues. My goal as a teacher is to explore both points of view using the specific elements of the case at hand—all the textual evidence that I marshaled above in discussing the pros and cons of Vermeule's belief about Miss Bates. The point of this activity is not to figure out who is right; it is to transform certainties generated by projection into questions, to eradicate knee-jerk reactions by restoring to students their talent for interpreting other people and themselves.

Who is right? Imagine that we were to sit down with Jane Austen and ask whether Miss Bates is angry. If she were to respond that both Miss Bates and she herself feel no resentment whatsoever about the condition of women in her society—that she is absolutely certain that neither of them is angry—that would give me pause. "The Lady doth protest too much, methinks" is a fundamental tenet of the folk psychology (picked up by Shakespeare, and uttered by Gertrude in *Hamlet*) that makes up interpretive talent: vehement declarations of certainty tend to be used as reassurance in the face of uncertainty. So, is Vermeule right about Miss Bates? We could count the bits of evidence for and against, and tally them on the board, in order to decide who is most right about Austen and Miss Bates, but the answer is the least interesting part of this conversation: most interesting is what happens to us on the way as we come to rest on a temporary conclusion, no matter what it is.

I say "temporary" because, if people are lucky, life is long, and what is a good answer now may not be a good answer later. In discussing ethnography as a "science without laws," Clifford Geertz (2007) describes the crucial commonsense, knee-jerk interpretive ability, fast thinking, as one that schizophrenics lack. Instead, they experience

> "the loss of natural self-evidence" (Blankenburg), the loss (now Sass) "of the usual commonsense orientation to reality, with its unquestioned sense of obviousness and its unproblematic background quality which allows a person to take for granted . . . the elements and dimensions of [the] shared world." This patient, a young woman, describes herself as without "the evidence of feelings," as lacking a "stable position or point of view" on life and how one ought to live it. She says everything strikes her as novel, "strange," "peculiar." She feels "outside," "beside," or "detached," as if "I was regarding from somewhere outside the whole movement of the world." "What am I missing, really?" she asks—"something small, [odd], something important, but without which one cannot live. . . . To exist is to have confidence in one's way of being . . . I need support in the most simple everyday matters. . . . It is . . . natural evidence which I lack." (222)

Geertz further argues that this estrangement from one's own habitual reasoning "is a general potential in human experience" that "ritual," which renders one's way of being "in a generically analyzable form," is designed to mitigate.[17] Similarly, when a life crisis—"the thousand natural shocks that flesh is heir to"—leads to questioning "one's way of being," literature can help us to be "beside oneself . . . in a sane sense" (Cavell 1981, 104). To think through a shock to one's own habitual interpretive strategies when everything seems "novel, strange," one might remember a novel, perhaps even the case of Miss Bates.

The aim of the discipline of statistics is to eliminate bias through numbers—regression analysis, sample size. Human interpretation does not suffer from "bias" in working toward truth; on the contrary, the point of interpreting literature is to discover the context of bias, and with it, various new possible forms of being, case by case. Insofar as they generate interest in cases and meanings, numbers can help, providing new cases or new interest in cases. What most interests me, however, is the potential for hermeneutics to inform numerical practices by providing methods for working through bias rather than simply attempting to eliminate it.

5.5. CULTURAL ANALYTICS

In the introduction to this volume, the editors quote William Davies's 2017 article "How Statistics Lost Their Power—and Why We Should Fear What Comes Next": Davies is worried about the current "shift from a logic of statistics to one of data" in which "a new, less visible elite" is employed by companies like FaceBook, "who seek out patterns from vast data banks, but rarely make any public pronouncements, let alone publish any evidence. These data analysts are often physicists or mathematicians, whose skills are not developed for the study of society at all." Earlier in my chapter, I mentioned that Merriman (2015) accuses Jockers and Moretti of ignoring the social sciences requirements to carefully construct a sample: only then would their results be generalizable to a larger population. Jockers argues in *Macroanalysis* that the bigger the sample, the more "correct" the results (2013, 6–7). No statistician would express such certainty about the validity of their work, as Johanna Drucker has noted (most recently 2020, 2, 177n1). More important, bigger datasets are not necessarily better, as Jockers imagines. In fact, Timnit Gebru was fired from the Ethics and AI team at Google for attempting to publish a paper in which she and Emily Bender argue that datasets which are too large can produce bad results— results that are both incorrect and unethical (Bender, Gebru et al. 2021).[18] How results are generated is a black box, to be sure, as Davies claims. Worse still, machine-learning models are generated *from data*, and when that data is gathered randomly in the wild, it teaches machine-learning algorithms to be as biased as the people who created that data through their clicks or tweets or blogs (Benjamin 2019; Noble 2018). However, there are methods for working through bias in a dataset that cannot be justifiably considered "a sample," and these methods come to us from sociology.

The first thing pointed out by the volume *What Is a Case?* (Ragin and Becker 1992) is that one can assume nothing about quantitative analysis without understanding the data:

> The view that quantitative researchers look at many cases, while qualitative researchers look at only one or a small number of cases, can be maintained only by allowing considerable slippage in what is meant by "case" . . . much of what is considered large-N research also must be seen as case-study research. (3–4; see also Gebru et al. 2020; So and Gebru 2020)

They demonstrate that even "big data" datasets are cases: cases of Twitter users, for example, who assuredly do not represent the total population. And, as it turns out, Jockers's dataset for his analysis of the difference between women's and men's writing is not reproducible on datasets created using the same constraints (Rybicki 2016): Jockers's novels are *a case*, the case of the 3,000-plus novels selected by Jockers.

Very far from seeing dataset construction and machine learning as mere fake news, however, we need to welcome the advent of massive amounts of digitized data as an opportunity. Building on sociology's efforts to articulate the kinds of claims that can be made based on cases, humanists have an opportunity to step into the fray and present our disciplinary methods for understanding "good" interpretations. At the forefront of such work, I would argue, are Ted Underwood, mentioned above, David Bamman, and Katherine Bode (all three working in the field of Computational Literary Studies), as well as mixed-method sociologists, including Laura Nelson, whose work is distinctive for her investment in the capacity of machine learning for intersectional research: the "unsettled categories" that bother Davies so much can help us see beyond the binaries that we impose upon data: m/f, white/non, and so on.[19] In my view, however, taking advantage of this opportunity will require interpreters of literature in the field of cultural analytics to scale down its promise to deliver empirical facts.

That work has already begun in earnest, with the publication of Andrew Piper's *Can We Be Wrong? The Problem of Textual Evidence in a Time of Data* (2020a), which carefully lays out how to render transparent the decisions made to generate quantitative results (56–60)—acts of interpretation, Drucker would call them (2020, 4–5)—and thus to adequately contextualize any generalizations based upon those results. And of course in literary studies, as Underwood (2013) repeatedly points out, there are questions worth asking that have empirical answers: when did novels start using the first-person perspective? When were certain genres most popular? (Don't like the way that "genre" is defined? As is *not* the case with monographic pronouncements, one can download the algorithm and dataset from the transparent cultural analyst and run a new algorithm that defines genre differently.[20] However, Piper's *Can We Be Wrong?* wants literary critics to jettison generalizations not derived from evidence—a good idea!—by becoming more sociological; doing so helps to "maintain our credibility as a discipline" (2020, 72). Like Jockers, Piper advocates for "moderat[ing]" [correcting?] generalizations made by literary critics "not through more hand-selected and often biased examples, but through the collection of larger and more independent [aka randomly collected?]

samples" (55-6). My interpolations inflect Piper's arguments in ways that he deliberately avoids, not thereby to set up a straw man, but to argue that "more" will not cure bias.

I would like to end by echoing a question with which Cavell begins his analysis of *King Lear*. Cavell was in part responding to the New Critics, arguably constituting one wave of the scientism that demands objective truth (Guillory 2002, 480, 497). Cavell had to begin by arguing that reading is not simply a matter of presenting facts, in this case, the facts of "the words on the page": "It is then unclear what the words are to be used as evidence for. For a correct interpretation? But what would an interpretation then be *of*?" (1969a, 268; emphasis added).

We currently live in a moment of "alternative facts": finding and stating facts based on substantiated evidence is crucially important in our time, but so is preparing people to hear them. We live in a moment of wild projections that fuel the conspiracy theories and hate rhetoric circulating madly on untrammeled communication systems. In addition to the transparent delivery of the facts, we need to teach people what it is like to think "as if" they were someone else and to view their own ideas from that perspective as a way of figuring out what is at stake in habitual ways of seeing the world. Northrop Frye wrote many years ago that literature offers the opportunity for "the recovery of projection" (1968, 14). I have argued that interpreting literary case histories cultivates projection in order to reveal bias, a task at which numbers can help—as long as we use them to read.

NOTES

This chapter was inspired by my sister asking me, "So what do you do when you read?"—she just wanted to know what experts do—and by the special issue of *PMLA* on "Cultures of Reading," edited by Evelyne Ender and Deidre Lynch (2018).

1. DH is a broad term indicating the use of computational methods to understand and preserve cultural materials. Within that field, "cultural analytics," "macro-analysis," and "computational literary studies" are primarily focused on quantitative modes of analysis (see Gavin 2020 for a discussion of these modes). Despite the angst over DH, quantitative approaches to literature precede it: "Quantitative interpretation of literature is a story that stretches back through book history, sociology, and linguistics to a range of nineteenth-century experiments. . . . This tradition is a branch of the digital humanities only in a parochial sense—as we might call pizza a branch of American cuisine. Both things were imported from a different social context, and inherit a longer history of their own" (Underwood 2017). Similarly, in *Enumerations*, Andrew Piper (2018, 1–2) argues that quantities of words have always been central to literary analyses.

2. See for example the special issue of *differences* called "In the Shadows of the Digital Humanities" (Weed and Rooney 2014), which sprang out of an MLA panel

called "The Dark Side of DH." Franco Moretti's work (2007, 2009a, 2009b, 2013) has been extensively discussed (Goodwin 2006) and attacked (Allington et al. 2016; Prendergast 2005; Trumpener 2009), but see especially Ross (2014) for background and bibliography. For an overview of Moretti's inauguration of "distant reading" and its relation to gender, see Klein (2016).

3. Articles in the University of Minnesota Press series of collections called Debates in the Digital Humanities, edited by Matthew Gold and Lauren Klein, confront issues surrounding the emerging field. Nan Z. Da recently published two articles (2019a, 2019b) that attack cultural analytics, the subfield of DH that uses numerical data from texts to support arguments about literature or literary history. The numerous responses are well worth exploring (Klein 2019; Piper 2020b; Schmidt 2019).

4. As Mary Morgan (2007) demonstrates, the case of the prisoner's dilemma overturns the cornerstone of laissez-faire economics—formulated first satirically in Bernard Mandeville's *The Fable of the Bees* (1714) and later in Adam Smith's *The Wealth of Nations* (1776)—that acting in one's own self-interest (acting "rationally," in economic terms) produces the best result for society as a whole. In history, Carlo Ginzburg (2007) argues, anecdotal evidence—documents representing the life of a Puritan entrepreneur living during the seventeenth century—seriously calls into question both Max Weber's *The Protestant Ethic and the Spirit of Capitalism* (1905) and Karl Marx's *Capital* (1867). Both are stories that have a dramatic impact on theory. Morgan's narrative changes in its details as it recurs in economic discussions, which makes it an "anecdote," as Paul Fleming (2012) defines it, while Ginzburg's anecdote depends upon the details—it constitutes what he calls "microhistory."

5. Stanley Cavell (1969a, 326–31) describes the "impotence to act" and react in relation to fictional characters imposed upon viewers by the medium of stage performance. I discuss the history of the notion of "aesthetic distance" and its emergence, in a specifically modern form, along with the advent of mass printing (Mandell 2015, chap. 2), focusing on Adam Smith's repeated equation of "sympathy" with "bringing the case home to ourselves" (Smith [1759] 1853, 13; repeated on 5, 14, 18, and 102).

6. Vermeule, though, would disagree with most of my argument because of its basis in psychoanalysis (2012, 427).

7. People tend to stop reading too soon, saying, "I disagree with the author about x," when it turns out that the author didn't say x. An important reading tool is what Gadamer calls "the prejudice of perfection" (1988, 76–78). Though you might stand back later and disagree, this activity requires believing that the author is right, temporarily, while searching to prove it in the author's terms.

8. Irony is taken to be the "Principle" of literature by Brooks (1979).

9. Jockers claimed in *Macroanalysis* to be analyzing literature with greater objectivity (2013, 6–7), but I am not sure that he would stand by that claim any longer. Regardless, even when they are attempting to conduct empirical studies of the broad outline of literary history, historical facts, the goal of the best work in the field is *not* to achieve greater objectivity than literary critics using traditional methods, only to achieve broadly true facts about literary history when it is possible to process texts computationally. For example, text-mining techniques can be used to ask and answer, "When did first-person narrative novels become prevalent, and are most novels still written in the first person?"

10. Statistics are visible on the right-hand sidebar here: https://www.hathitrust .org/about.

11. I am grateful to Philip Galanter for this insight.

12. Because the form of topic modeling called LDA is based upon Bayesian probability, it is not possible to produce the same results each time one runs it. Further, removing from the text character names and "stopwords" ("the," "a," "mr," for example) radically alters the results. The top thirty words constituting the "topic" containing "obliged" and "obliging" from my run of *Emma,* with character names and stopwords removed and 100 iterations used, are:

> soon come say sure wish came pleasure away word obliged old told hear comfort looking looked way ill just giving kind smile best saying obliging honour ashamed regret round knowledge

For explanations of topic modeling, see Jockers (2011) and Merriman (2015). Mimno (2013) developed the Topic Modeling tool that has been incorporated into Voyant (Sinclair and Rockwell n.d.).

13. A generous reader for this volume recommended Lorraine Daston's notion of "reasoning from exemplars" (2016). According to this reader—I could not have said it as well—Daston argues for "cultivating the capacity for a form of judgment not reducible to rigid rules" (Reader 2).

14. I am happy to supply details about the modifications and the version of the stripped-down text that I used (mandell@tamu.edu).

15. See, for example, the numerous articles in the supplement to the journal *The American Statistician*, "Statistical Inference in the 21st Century: A World Beyond p < 0.05," https://www.tandfonline.com/toc/utas20/73/sup1?nav=tocList.

16. Note that the analyst is emphatically *not* the artist here: "No one wants to be someone else's work of art" (Schwaber 1983; Treurniet 1997).

17. Geertz (2007) quoting Sass (2004, 305–6), who quotes and translates Wolfgang Blankenburg (1971). Geertz modified the translation of *komisch* to "odd"; Sass had translated it as "funny," as Geertz explains (224n21).

18. For discussions of Gebru's firing by Google, see Hao (2020) and Simonite (2021). While the latter is worth reading and may be well intentioned, Simonite's article could provide fodder for casting Gebru as a psychologically scarred troublemaker, a possible reading with which I strongly disagree. It is easier, but unproductive, to see protests against micro- (and in this case macro-) aggressions as generated by personal psychological problems.

19. Bode's (2020) review of Underwood (2019) and Nelson's recent talk at Texas A&M University (Nelson 2020) are especially relevant to this chapter. I am referring to Bamman's work more generally (https://people.ischool.berkeley.edu/~dbamman/).

20. Ted Underwood has made this point in presentations for the NovelTM group.

When Bad Numbers Have Good Social Effects

CHAPTER 6

Why *Five* Fruit and Veg a Day?

Communicating, Deceiving, and Manipulating with Numbers

Stephen John

Our world is full of numerical measures, targets, and indicators, ranging from blood-pressure readings to economic growth targets to standardized test scores. The notion of being for or against all forms of quantification is as nonsensical as the notion of being for or against all language. Despite the heterogeneity of quantitative measures, however, we can distinguish two main purposes of quantification: as a way of gaining a better understanding of some natural or social phenomenon and as a tool for changing behavior. For example, we might measure neonatal mortality rates as a way of understanding the quality of care provided by a maternity ward or as part of a program of naming and shaming "bad" midwives. In turn, we can distinguish two ways of criticizing some quantitative measure. First, that it does not facilitate understanding of the relevant phenomenon: for example, neonatal mortality statistics may not be a useful guide to quality of care if they fail to take account of whether negative outcomes were avoidable (O'Neill et al. 2008). Second, that its use is socially or ethically or politically problematic: for example, using outcome measures to name and shame bad midwives may incentivize midwives to refuse difficult cases, with bad effects for mothers and children. While it doesn't make sense to be for or against quantification per se, it does make sense to be against some quantitative measure used in some context because it is epistemically problematic, because it is ethically or politically problematic, or both.

These comments are very abstract and general. There is much more to be said about the different ways in which we might criticize numerical measures on epistemological grounds, ranging from (relatively) simple concerns about accuracy to concerns that quantitative measures can never capture the affective dimensions of experience. Similarly, there is much more to be said about how to think about the ethics and politics of measurement; consider, for example, the difference between criticizing a measurement scheme in terms of its putative aims and in terms of its actual consequences. Moreover, the relationship between broadly epistemic and broadly practical critiques is itself tricky. On the one hand, we clearly can—and often should—separate epistemic and ethical concerns: neonatal mortality might tell us a lot about quality of care, even if we have excellent reasons not to shame midwives; conversely, we could have good reasons to use neonatal mortality rates as a target for midwives—say, because they create a sense of shared mission, even if we think they are pretty useless indicators of care quality. On the other hand, these two sorts of concerns are often intertwined. The use of epistemically misleading indicators is often ethically problematic: using epistemically misleading indicators to name and shame midwives is not only misleading but unfair. Conversely, using a quantitative measure as a target can undermine its epistemic usefulness; for example, neonatal mortality rates might be a decent measure of quality of care at some point in time but their use as targets might lead people to game the system (Bevan and Hood 2006).

This complex relationship between broadly epistemic and broadly ethical concerns is not unique to the case of numerical indicators but familiar from philosophical thinking about the ethics of deceit. On the one hand, there is a familiar, everyday distinction between the question of whether some claim is *true* and whether it is *good* or *right* to say it; it's cruel to say that someone's new haircut is horrible, even if it is. However, this distinction becomes complex in cases of deceit: a lie is *ethically* wrong (in part) because it is an *epistemically flawed* representation of the world (see Carson 2010 for useful discussion). In turn, we often use the language of deceit to talk about tools of quantification; there are "lies, damned lies, and statistics." My aim in this chapter, then, is to ask how the older tradition of the ethics of deceit can help us think through our more recent task of understanding the limits of the numerical.

Specifically, drawing on the long literature on deceit in medical ethics (Jackson 1991), I consider the use of quantification in the domain of health care and health advice. Of course, quantified claims about medicine and health are generated and communicated in many ways and for many reasons. Some of these cases have already been extensively discussed—for example, how the use of "framing effects" in presenting probabilistic data may undermine consent (Chwang 2016). As such, a full account of deceit and quantification in health care is beyond the scope of this chapter. Rather, I will focus on one particularly interesting, underexplored phenomenon: the communication of "spuriously precise" numbers, such as the dietary advice that we should each eat five portions of fruit and vegetables a day. What may seem a distinctively modern phenomenon—the promulgation of quantified targets for self-improvement—may be conceptualized using older philosophical tools.

Section 6.1 outlines the concept of "spurious precision," using the "five-a-day" case as an example. Sections 6.2 and 6.3 then discuss the ethical issues arising in this specific case. In section 6.4 I build on this analysis to develop three broader lessons for the ethics of spurious precision. In the conclusion I sketch some of the implications for the even broader task of developing an ethics of deceitful uses of numbers in general, and hence the relationship between different perspectives on quantification.

6.1. SPURIOUS PRECISION:
UNCERTAINTY AND VAGUENESS

Health apps state that the average person should walk at least 10,000 steps a day; in the United Kingdom, a body mass index (BMI) measure of 25 marks the difference between being a "healthy" weight and being "overweight." These very different numbers have two things in common. First, they can have important consequences for individuals; for example, some companies provide fitness trackers to employees, rewarding them for meeting the 10,000 steps target (O'Carroll 2018), and patients' entitlements to treatment can differ depending on whether their BMI score is over or under 25. Second, other, equally precise numbers would also be justifiable. We have excellent evidence that physical activity improves physical and mental health. However, no one seriously thinks that walking 10,000 steps is privileged over 9,900 or 10,100 steps. Similarly, although it is uncontroversial that someone with a BMI of 15 is not overweight and someone with a BMI of 35 is overweight, there is no straightforward reason to treat 25 as the key cutoff point for being overweight. (Indeed, in the United States, the cutoff point is set at 30.) In both the 10,000 steps and BMI cases we are dealing with what I call "spuriously precise" numbers: numbers that are justifiable, given our evidence, theories, and conceptual frameworks, but where other numbers would be equally justifiable. The rest of this section develops a more detailed account of this phenomenon with reference to the health advice familiar to anyone living in, or merely visiting, the United Kingdom, that he or she should eat five portions of fruit and vegetables a day.

The five-a-day advice, often accompanied by guides as to what counts as a "portion" (a handful of peas, a single orange, two small plums, and so on), has moved from official leaflets through advertising campaigns to school classrooms to become an everyday trope, with playground jokes about fruit-flavored sweets or Jammy Dodger cookies counting as "one of your five a day." The UK campaign was started in 2003, based on a finding in a World Health Organization (WHO) report that, for adults, eating at least 400g of fruit and vegetables a day reduces the risks of atherosclerosis and coronary heart disease, as well as many cancers (Batty 2003; World Health Organization 2003). (The UK government assumes a portion is 80g.) Many other governments have also adopted public health campaigns based on the WHO's work. However, their recommendations differ widely: while the German government advises five portions, the Canadian government advises 7–8 portions for women and 9–10 for men; the Austrian government advises five portions but of 200g each; meanwhile, in Singapore, the public is told to try a daily 2 + 2 diet (two portions of fruit and two of vegetable) (Peters 2014).

We can interpret each of these campaigns as resting on or implying a factual claim—that a certain level of fruit-and-vegetable consumption would promote individual health—but as disagreeing on the precise level of consumption required for a healthy diet. In the face of such variation, there are two options. One is to think that one of these divergent targets must be best, in the sense of most accurately representing the "true" answer: how much fruit and veg is *actually* necessary for good health. The alternative is to think that, given the nature of the claim in question, each recommendation may be defensible. In this case, the second option seems preferable for two reasons: claims about healthy eating are subject both to epistemic underdetermination and objective vagueness worries.

First, the evidence behind claims about dietary consumption is hard to collect and to interpret. Given variation between individuals, the huge range of determinants of individual health, and the difficulty of conducting large-scale longitudinal studies, it is difficult to pinpoint with any accuracy the effect of a single dietary change on a range of health outcomes (Jukola 2019). We should not be surprised that in 2017 alone, a news story in February summarized a new study as "Eat 10 fruit-and-veg a day for a longer life, not five" (Telegraph Reporters 2017), but by August another study was reported as "three portions of raw fruit-and-veg as healthy as five a day" (Smyth 2017). Note that this is not to say that *any* claim is epistemically respectable. There are good reasons to think that, in general, increased consumption of fruit and vegetables is beneficial for our health, such that, for example, a "two a day" message would not be compatible with our evidence, whereas a "twenty a day" message would be overkill. It is, however, to say that claims about the effects of dietary changes on outcome measures are *epistemically underdetermined*, in the sense that more than one estimate is reasonable given our evidence. In turn, it is a platitude in contemporary philosophy of science that our nonepistemic interests and values may play a legitimate role in responding to problems of underdetermination (Biddle 2013), leading to defensible differences between different estimates based on the same evidence.

Second, the concept of "health" is itself vague. Consider the analogy with baldness. Clearly, some people are bald and other people have luxuriant hair. Someone with ten hairs on his head is bald, someone with 10,000 hairs is not. However, it also seems there is no simple matter of fact about the number of hairs you must have on your head if you are not to be bald: nothing in nature means that 5,000 hairs would mark off "baldness" from "nonbaldness," in a way 5,001 hairs would not. Therefore, in the unlikely event we had to specify the *precise* numerical definition of baldness, our choice within a certain range would be determined by *practical* concerns, rather than the way the world is. (Strictly, some philosophers disagree with this account of vagueness—see Williamson 1994—but even they agree that, for practical purposes, in making precise vague concepts we often have to simply choose an essentially arbitrary number within some nonarbitrary range.)

Consider now the case of health. Many diseases are characterized in quantitative terms; for example, some forms of heart disease are defined in terms of blood pressure. Clearly, in such cases, deciding on any specific blood pressure as marking the boundary between diseased and healthy seems, like the question of when one is bald, to involve an essential vagueness (Rogers and Walker 2017). Therefore, even if we had perfect knowledge of how precisely fruit-and-veg consumption affects our bodies, we might still reasonably disagree on what constitutes health. Note, again, that this does not mean anything goes. It would be wrong to say that someone with very low or very high blood pressure is healthy. Rather, my claim is that there is more than one reasonable way of making our everyday qualitative concepts, such as health, more precise. In this regard, the five-a-day message resembles the problem of setting a BMI level for being overweight; although some thresholds or measures are clearly incorrect, more than one threshold or measure can be justified.

In some cases, we disagree over precise numbers, where—in principle at least—one of us must be right and the other wrong: there is a fact about your blood pressure, even if your monitor and my monitor give different readings. We should not think that the disagreement between various public health experts in various nations over dietary advice is like those cases. Given our evidence, it is plausible to say that, in general, eating several portions of fruit and vegetables a day improves the average person's health. However, given that the case involves *both* epistemic underdetermination and objective vagueness, more precise claims—for example, that eating five portions is particularly important—are spuriously precise; other, equally precise estimates, are also defensible. Spuriously precise numbers are epistemically justifiable in one sense—they fall within the plausible range of estimates—and, yet, epistemically problematic in that they disguise a more complex, ambiguous picture.

6.2. SPURIOUS PRECISION: DECEIT AND MISLEADING

The previous section sketched an account of spurious precision. In this section and the next I turn to consider the ethics of spurious precision. We use spuriously precise numbers for a wide variety of reasons. Having a nice, simple, round target of 10,000 steps may be an effective way of encouraging oneself to exercise more; for practical purposes of patient management, there needs to be some cutoff point to decide when patients are overweight, and a BMI of 25 might be a sensible point to choose. My interest in the rest of this chapter is with one specific kind of use, exemplified by the five-a-day campaign: where experts communicate spuriously precise numbers to audiences. When we think about numbers as embedded in communicative actions—rather than as tools we use to motivate ourselves or to guide clinical practice, for example—we can think about their use in terms of the long, rich tradition studying the ethics of deceit. In this section, I develop an argument that, from this perspective, the five-a-day target may seem particularly problematic.

The ethics of deceit has a long history in traditional medical ethics; for example, in cases where a physician hides a "distressing" diagnosis from a vulnerable patient or where a research subject is given a placebo. Is the five-a-day message deceitful? Paradigm cases of deceit involve two elements: first, that a speaker misrepresents her beliefs about the world; and second, that she does so to secure some further, nonepistemic goal which follows from her audience believing the misrepresentation (John 2021). Clearly, we can deceive using quantitative measures; for example, imagine a physician who knows that a patient will change her diet if she is told that her blood pressure is high, so he reports a false blood-pressure reading to engineer this behavior change. Note that both conditions are necessary for a speech act to count as deceitful: a physician who reports a false blood-pressure reading out of sheer ineptitude fails in her professional obligations, but is not deceitful. Conversely, even if a physician reports a patient's blood pressure to him with the intention of changing his behavior, her claim is not deceitful if the estimate is accurate. (Note that there is some disagreement over the relationship between the two elements listed here—see, for example, Carson 2010. For our purposes, it suffices to say that deceit normally involves both components.) In turn, there is wide-spread consensus that deceit is, at least prima facie, ethically impermissible, regardless of whether the consequences are valuable. For example, most authors suggest that even if the physician's deceit leads the patient to make positive changes to her diet, it remains ethically wrong (Jackson 1991). Such views are particularly strongly held in bioethical debates, where they are typically discussed as instances of autonomy-undercutting paternalism (Beauchamp and Childress 2001). Interestingly, however, even writers who are skeptical of popular conceptions of autonomy condemn deceit (Manson and O'Neill 2007).

The claim that we should eat five a day is true, in the sense that eating this amount of fruit and vegetables would, for most people, improve their health (given that average consumption falls well below this level). Hence, it is not a lie. However, a communicative exchange can be deceitful even if the speaker does not say anything straightforwardly false (Saul 2012). Imagine a doctor who reports to a patient that his blood tests show no evidence of cancer, while failing to mention that the X-ray results showed a significant tumor in his lung. If the blood tests are clear, the doctor has not lied to her patient. However, she has deceived him, because she can (or should) foresee that, on the basis of her true claim about the blood tests, the patient will infer a false claim: that he does not have cancer. The patient will draw this conclusion because he assumes that a doctor will obey what Grice (1975) called "conversational maxims," informal rules which structure ongoing communication. In the case at hand, the physician violates the maxim of *quantity*—that, in a cooperative conversational exchange, a speaker will offer as much information as she possesses that is relevant to the hearer within the context of that exchange. When the doctor offers only the blood-test results, the patient infers that these are the only relevant pieces of information the doctor possesses. Although nothing the doctor has said is strictly and literally false, her silence implies a falsehood.

With this background, let us turn back to the five-a-day campaign. Although it makes no straightforwardly false claim, the choice to report a specific number implies that a particular number is special—that five somehow marks a particularly important level of consumption. Again, we can think of this implication in terms of the maxim of quantity; by asserting a specific number, rather than a wider range, the speaker implies this number is certain and precise, when in fact it is a generalized approximation. Therefore, when the speaker chooses to assert the spuriously precise estimate, rather than the range, she misleads the speaker in making her think that this precise recommendation—though reasonable—is scientifically privileged over other precise numbers in a way that it is not.

Furthermore, this claim is reported precisely because policymakers wish to change behavior, namely the hearer's eating habits. Therefore, even if it is epistemically reasonable, the five-a-day message seems designed to mislead, and is therefore deceitful.

There is much disagreement over whether misleading speech is as ethically problematic as downright lying. Following Kant, some hold that lies are always impermissible, but other forms of deceit are sometimes permissible (Kant 1997). Others deny that this distinction bears much moral weight or, indeed, that it can easily be drawn at all (Saul 2012). Most arguments for treating the act of misleading as less problematic than lying assume that careful hearers might be able to challenge misleading speakers in a way that they cannot challenge downright liars. Consider the patient who asks, "Are those the only tests you did, doctor?" (Webber 2013). As such, a standard argument runs, misled hearers are implicated in their own misunderstanding by failing to demand further information. Hearers who are simply lied to, by contrast, do not share in the responsibility for forming false beliefs. The speaker alone is to blame.

There are good reasons to be suspicious of such arguments (Rees 2014). Even if they are defensible in the abstract, their applicability in specific cases presupposes two factors: first, that there is some degree of epistemic symmetry between speaker and hearer; and second, that the hearer has some reason to be suspicious of the speaker. When neither of these conditions holds, misleading speech can be considered equivalent to outright lies. That seems to be the case when government experts issue recommendations to the public. Experts are epistemically superior to nonexpert hearers (at least in some relevant domain), such that it is unreasonable to expect nonexperts to challenge the experts' pronouncements. Furthermore, when we consider the speech of government agents, such as public health officials, we are far from the paradigm examples of cases where hearers are or should be suspicious, such as courts of law or used-car showrooms. Therefore, the five-a-day claim misleads in a way that is ethically equivalent to a lie.

At this point, it is useful to note an important difference between the five-a-day case and other cases of spurious precision. Consider a scientist who offers a policymaker some precise figure of the likelihood of some future climate event, without mentioning that this is an estimate, and far from certain. On the face of it, this kind of case resembles the five-a-day case. Yet, as I have argued elsewhere (John 2018), such apparently dishonest speech may be morally irreproachable. That is because, given the huge confusions and controversy around climate science, it may be the only way for the scientist to effectively communicate the well-established fact that climate change poses a serious threat to humanity. The spuriously precise claim is a vehicle for transmitting a more general message; it is "well-leading" rather than "mis-leading." The scientist's speech resembles the teacher who deliberately obfuscates and ignores various complexities as a way of conveying a general understanding to her students. Unfortunately, we cannot justify the five-a-day case as an instance of such "epistemic paternalism" (Ahlstrom-Vij 2013), because the relevant well-established claim—that *more* fruit and veg is *better*—could just as easily be conveyed without making a spuriously precise claim. The point of asserting the precise claim is not to convey some information more effectively because the relevant informational content can be communicated more accurately.

Apart from epistemic reasons, why offer precise, rather than vague, estimates? In our case, there were two messages that could be promoted. One was something along the lines that "eating more fruit and veg is, in general, a good idea"; the other was "eat five fruits or vegetables a day." Presumably officials make the second, spuriously precise, recommendation on the assumption that citizens are more likely to change their behavior in response to targets that are simple and apparently obtainable rather than complex and vaguely informative. The simplified message plays on (arguably) subrational features of our psychology, such as the strong incentive provided by the felt reward of achieving some target. As such, the decision to report a spuriously precise estimate seems not only deceptive but manipulative, an attempt to influence a hearer's beliefs or behavior in a way that bypasses her rational thought, where this influence is opaque to the hearer (Sunstein 2014). Note that not all cases of deceit are manipulative. A physician who invents some blood-pressure reading, with the aim of making her patient take his statins, deceives her patient but does not manipulate him. In our case, though, the point of the deceit is not merely to change behavior, but to do so in a particularly underhand way, by playing on our unconscious, psychological reward system.

6.3. SPURIOUS PRECISION: MANIPULATION

Normally, to say that some act of communication is both misleading and manipulative is a serious charge. The charge seems even weightier in the case of official government advice. On the other hand, the five-a-day campaign does not seem problematic at first glance—it is hard to imagine shock and outrage at anything I have said above. Why should that be? In this section, I sketch some reasons why, even if manipulative deceit is usually problematic, this case may seem less problematic, and assess the weight of those reasons. The subsequent sections suggest some more general lessons for thinking about the use of numbers.

Just how serious is the charge of manipulation? None of us is a rational automaton; we all have biases and psychological quirks. We might use these quirks in all sorts of ways to navigate our lives, including through the use of spuriously precise numbers. For example, we might set ourselves numerical goals that are ultimately arbitrary—"today, I will write 1,000 words" or "by Christmas, I'll be able to run 5 kilometers nonstop"—as a way of ensuring we achieve our more general aims of finishing a book or improving our stamina. We may be manipulating ourselves, but it would be very strange to condemn all such cases as unethical. Relatedly, we often explicitly or implicitly allow others to manipulate us. In general, such cases are unproblematic. Imagine I employ a personal trainer to help me lose weight: if she tells me I *must* run 5 kilometers nonstop by Christmas, appealing to my reward mechanisms, she is manipulating me, but such manipulation is clearly fine—it's what I paid for!

Can we, then, justify the manipulative deceit of the five-a-day message? To be sure, there is no explicit contract in this case that would be analogous to my contract with the personal trainer. However, we might think there is an analogy with cases where a physician misleads a patient about a drug's composition because she wishes to engineer a placebo effect (Lichtenberg et al. 2004). We cannot seek *actual* consent for administering placebos because doing so would undermine the effect. However, at least in some cases, we can be fairly certain that patients *would* consent to being deceived and manipulated in this manner. One might think something similar is true in the five-a-day case: that citizens can be presumed to give a kind of "hypothetical consent" to being manipulated in this way.

To make this analogy more vivid, consider a mundane, nonquantitative case: the labeling of defibrillators. Defibrillators are used to administer electrical shocks to patients who enter cardiac arrest, via two pads, one placed on a patient's left side, one on her right. Increasingly, they are placed in public areas, allowing nonmedics to respond to emergencies. Typically, pads are marked "left" and "right," indicating the body of the person receiving the treatment. Strictly speaking, however, these labels are unnecessary, because the pads are interchangeable. Why, then, are the pads labeled at all? A standard story runs as follows: when pads are not labeled, users who are not medically trained tend to panic, worrying about administering the treatment wrong. When the pads are labeled, this worry is allayed, as there can be no uncertainty. Assuming this story is correct, we have a (peculiar) analogue to spurious precision: although it is true that a "left" pad could be placed on the left, marking it as such implies that this placement is privileged, when in fact it is not. This communicative choice assumes that hearers are incapable of complex thought,

but need to be manipulated into acting properly. Yet this practice seems
not only justifiable, but commendable. Why? When I imagine a situation
where I need to administer defibrillation to someone, I can (all too easily)
see myself panicking at the question of which paddle to place where. In
those circumstances, an instruction—however unnecessary—would help
me pick up the defibrillator in the first place and achieve my goal of helping someone in need. Given a realistic estimate of the limits of my own
rationality, I would consent to being manipulated in this way, and it seems
plausible that each one of us would consent to the deception in question.
In fact, anyone who objected would seem rather weird. As such, it seems
ethically unproblematic—indeed, considering the worry allayed and the
lives saved, ethically praiseworthy—for manufacturers to label the pads.

Much public health advice—including the five-a-day campaign—makes
a similar assumption: that everyone wants to be healthy, such that it does
not matter whether claims are deceitful or manipulative, as long as making
them tends to lead to healthy behavior. On this model, the number five being arbitrary does not undermine the propriety of insisting that everyone
should eat five a day, as long as that recommendation has the positive consequence that people eat *more* fruit and vegetables. Can we appeal to this kind
of argument to justify the manipulation? Clearly, there is something to this
reasoning. Arguably, the state has legitimate reasons to encourage its citizens to live healthily, and most of us broadly welcome interventions to that
end. Indeed, an interesting feature of the five-a-day case is precisely that,
despite constant bombardment, it is unclear whether anyone really believes
that five is a magic number. As I noted above, in everyday life, people often
talk about their five-a-day as something that is a bit silly but also worth re-
membering. Even if the advice is deceitful—in making it seem as if five is a
magic number—it's not clear that anyone is *actually* deceived, that anyone
actually forms the false belief that exactly five portions are required for being healthy. Instead, they *act* as if they were deceived, by using the number
as a useful heuristic to simplify complex choices. This description of how
the public is affected by the five-a-day message points to a more general feature of precise recommendations: that they can serve an important role in
guiding and coordinating action even when their audience knows them to
be spurious. In this case, then, there is an interesting gap between our epistemological concerns (that the statement falsely implies that a certain number is special) and our practical interest in simplifying everyday life. What
looks like a deceitful way of representing the world is more like an implicitly agreed-upon strategy for one party to help the other achieve some goal.

Still, even with these remarks in place, there are two grounds for
concern. First, it is not clear that this message does have the positive

consequences its proponents claim. From a narrow, instrumentalist perspective, the effects of nutrition campaigns on consumption are contestable (Capacci et al. 2012). From a broader perspective, even if the five-a-day message has some positive effects, it also has negative consequences, by placing the onus for healthy eating on individual agents rather than addressing the larger structural factors that affect both access to food and attitudes toward it. If we start from the assumption that manipulation is wrong by default and requires a positive consequence in order to be justified, it's not clear what we should make of the five-a-day campaign. Second, and more fundamentally, we should question the assumption that each would consent to her dietary choices being engineered for the sake of promoting her health. To make this assumption is to overlook the complex ways in which the value of health plays into the broader concept of living well, and the social and cultural roles of consumption within a well-lived life. It may well be that most members of society value their health— understood to include the reduction of risk factors for disease—to such an extent as to license government intervention in their broader consumption choices. However, it is far less clear that these values are shared by all of society. Spuriously precise numbers may be consented to by *most* of us, but not necessarily by *all* of us.

Deceit and manipulation are normally bad. However, they are not always bad: the personal trainer *should* manipulate me; labeling the defibrillator pads is *helpful*. As suggested above, we might justify spurious precision by appeal to hypothetical consent. Just as, given the practical stakes, reasonable people would not oppose the labeling of defibrillator pads, so too reasonable people would not oppose spuriously precise recommendations. However, the two cases are not necessarily parallel. That is because in picking up the defibrillator, we are committing ourselves to a goal—helping someone—and the labeling is a means to help us achieve that goal. Using the spuriously precise dietary advice, by contrast, is not simply a way of helping us achieve our goals, but involves making assumptions about what we want and what we value, and, in turn, implicit judgments about what we ought to want and ought to value.

6.4. THREE LESSONS ABOUT THE ETHICS
OF SPURIOUS PRECISION

In section 6.1, I outlined the notion of spurious precision, as exemplified by the five-a-day message. Section 6.2 argued that this message is a case of manipulative deceit. Normally, we think that such deceit is wrong, but as section 6.3 argued, the case is more complex—there are reasons for and against this spuriously precise recommendation. Rather than engage in more casuistry around this specific case, in this section I suggest three more general lessons for thinking about the communication of spuriously precise numbers: first, that there can be no argument for or against the use of spuriously precise numbers per se; second, that the decision whether to use spuriously precise numbers requires deliberating about some distinctively political issues; and third, that we need to pay particular attention to how spuriously precise numbers travel.

One might think that *any* spuriously precise number is, in a straightforward sense, epistemically problematic, and hence that its use in private or public life is problematic. A running theme throughout my analysis, however, has been that regardless of their epistemic status, spuriously precise numbers can play valuable roles, both for individuals (say, motivating healthy eating) and for societies (say, allowing for efficient categorization or helping to organize collective action). It may seem, however, that my discussion of deceit implies a narrower "primacy of the epistemic": that we should not *communicate* spuriously precise numbers to others, because to do so is deceitful. Again, however, my arguments have revealed a more complex picture. The propriety of speech acts depends, in part, on the social role of the speaker and associated social expectations; it may be unproblematic for a personal trainer to insist on five a day but problematic for a scientific expert to do the same. Personal trainers are supposed to say whatever is helpful for us to achieve some goal, whereas scientific experts are supposed to inform us of the facts. It is in this latter, "informing" role that spurious precision seems problematic. (For more on the notion of an informant, see Manson and O'Neill 2007). However, as I suggested above, it can be unclear whether we should think of public health advice as akin to the words of an informant or a trainer. Because social roles and expectations are fluid, and intentions hard to disentangle, it will often be difficult to say that some particular speech act was deceitful or manipulative. Here, then, is the first conclusion: there is no general argument for or against the communication of spuriously precise numbers as such.

The notion of hypothetical consent links spurious precision to the literature around "libertarian paternalism" or "nudging" (Sunstein 2014). Proponents of nudging argue that we should exploit irrationalities in human psychology to ensure that people make choices that are in their best interest or those of others. For example, we can use our knowledge of optical illusions to paint road markings in such a way that drivers slow down when taking a bend. One way of justifying such manipulation is by appeal to (a version of) the placebo argument: we would agree to these forms of manipulation were we aware of our own limitations. (Or, at least, it seems that we *should* consent to a placebo; of course, whether we would consent is complicated because we might be reluctant to acknowledge our irrationality.) I have argued that we might think about some cases of spurious precision in an analogous way: as justifiable when they use our cognitive biases—in this case, our bias toward clean, simple numbers—as a way of nudging our behavior to the good.

Unfortunately, not all cases of nudging are as benign as the road markings. Rather, similar techniques may be used for purposes that crowd out rational deliberation on prudentially, ethically, or politically complex and sensitive matters—consider how nudging techniques might be used in political campaigning. Similarly, not all cases of spurious precision are as benign as the defibrillator; when choices are prudentially or ethically complex, we should be enabled to make them autonomously, in ways which reflect our values, rather than have others decide for us, precluding our own deliberation from the process. Still, we should be careful in our thinking about spurious precision not to idealize notions of autonomy, rationality, and understanding. Precisely because we are cognitively limited agents, we might all welcome the help of simple, perhaps manipulative numbers in the decisions we make in some areas of our lives. Insisting that spuriously precise numbers are always wrong because they play on our irrationalities is like stamping one's foot at the deceit involved in the road markings. What is tricky is finding the balance between cases where manipulation is sensible or praiseworthy and cases where it undermines rather than promotes agency. Debates over communicating numbers are not, then, merely ethically complex but also shot through with larger questions about the limits of permissible state action, in particular the extent to which we can reasonably assume that audiences do, might, or should agree to be manipulated. Even if you disagree with my analysis of the five-a-day case, you probably disagree because you have concerns about the substantive ethical and political issues above, rather than because you think spuriously precise numbers should *never* be communicated. Similarly, if you wonder whether we should be as worried about the use of spuriously precise numbers by private companies as by government officials, your concern has less to do with numbers and more to do with the structures of the market economy. Here, then, is my second general conclusion: any assessment of the communication of spuriously precise numbers is intimately linked to structural ethical and political questions.

In arguing for these claims, I have sought to domesticate the notion of quantification and to treat messages involving quantified measures as just another kind of statement; that is to say, I have argued that the ethics of using numbers can be understood as part of the ethics of communication, rather than as a separate topic. It is important, then, to stress that there are differences between quantified targets and the kinds of qualitative statements that we normally think of when we talk about deceit or manipulation. Some of these differences serve to heighten familiar tensions. For example, many of us are aware of our own innumeracy, such that complicated numerical claims may be daunting. Therefore, drawing a sharp ethical distinction between lies and other forms of misleading may be particularly difficult in cases of spurious precision. Relatedly, we tend to associate precise numerical measures with forms of objectivity, such that we might be less likely to recognize that numbers can hide conceptual ambiguities and uncertainty, increasing the risk that we will be misled by their use. "Spuriously precise" numbers may be a particularly useful—and murky—way in which to deceive others.

However, perhaps the most important feature of numerical measures, and one that complicates the ethics of spurious precision, is their ability to travel. To explain this remark, consider one last time my central example. The five-a-day advice is ubiquitous: we find it everywhere from leaflets in GPs' clinics to adverts in supermarket aisles. In this regard, the message exemplifies a very general feature of numerical claims: that they often "travel" from their original setting to a wide range of other contexts. In light of such complexity, it doesn't make sense to say that each and every utterance of the five-a-day claim is deceitful or manipulative, because it is not always made by someone perceived to be an expert and so does not necessarily enjoy the aura of authority.

If I have been taken in by a lie and repeat it, then my action is not ethically reproachable but it still seems epistemically problematic, because a falsehood becomes further embedded in our communal life. Something similar is true in the case of the constant repetition of the five-a-day message: repetitions of that claim may not themselves be deceitful or manipulative, but each serves to further embed the claim in public discourse and consciousness. Unlike the lie, however, the claim at hand is not straightforwardly false. Rather, the problem with proliferating it is that doing so perpetuates a different falsehood: that we can possess certainty and conceptual clarity about the relationships between diet and health. In turn, this falsehood perpetuates a cycle whereby the public demands and researchers promise simple answers to complex questions. Even if no one is entirely taken in by the five-a-day message, the circulation of the claim gives credence to a misleading sense that the epistemic, conceptual, and ethical ambiguities inherent in relating diet and health *could* give way to simple, numerical claims.

The fact that numbers travel so easily complicates assessment of my case. It also implies yet another important complexity for the ethics of spurious precision in general. Above, I stressed that the ethics of communicating spuriously precise numbers is complex; such numbers need not be deceitful or manipulative; even when they are, the relevant forms of deceit or manipulation might be justifiable. However, whenever we use a "spuriously precise" number, there is always the possibility that this number will be transplanted from the initial context—perhaps a context where all involved are aware of the number's limitations—to a different context, where such awareness is absent. In turn, the use of spuriously precise numbers in these new settings can lead to errors or misunderstandings, with potentially significant consequences. Even when the use of some spuriously precise number seems benign in a given context, we may have a special responsibility to consider what will happen if and when that number travels to new contexts. Here, then, is my third and final general lesson: if spuriously precise numbers cannot be controlled, then we may have reasons not to use them in the first place.

6.5. CONCLUSIONS

This chapter started with a very general sketch of two perspectives we might take on numerical measures, indicators or targets: in terms of their epistemic adequacy and in terms of their ethical, social, or political consequences. However, for most of the chapter I have focused on a single problem: how to think about the ethics of communicating "spuriously precise" numbers. In conclusion, I sketch some of the ways in which this narrower question relates to the broader theme of different ways of criticizing and justifying quantitative measures.

Very often, we communicate numerical measures or targets to others because we wish to change their behavior. Sometimes it is permissible or proper to seek such change, and sometimes not. However, a very general condition on any communicative action is that it not be deceitful. I have examined one form of deceit—in the form of spurious numbers—but there are many other ways in which numbers might deceive. In the 2008 US presidential election, for instance, Rudy Giuliani made much of the fact that five-year cancer survival rates are higher in the United States than in the United Kingdom, with the implication that socialized health care is less effective than private care. However, as Gerd Gigerenzer points out, five-year survival rates may be a poor guide to overall rates of cancer morbidity and mortality, because they can easily be improved by identifying cancers at earlier stages, obviating the need for treatment (Gigerenzer and Wegwarth, 2013). By Grice's communicative maxims, Giuliani's true claim implied a false claim. One general task for an ethics of deceit fit for the modern world is to understand how and when numbers can deceive. This task is complicated because a fear of complex statistics may heighten the epistemic asymmetries which underlie deceit, because numerical measures may seem objective in ways that preclude challenge and criticism, and because numbers have a tendency to travel into new contexts. The statistics that Giuliani used may have been collected for excellent reasons and been highly illuminating in their original context, but once they were moved out of that context, they took on their own troubling life.

The practical payoffs in developing such an ethics are obvious. There is also a more general theoretical payoff for thinking about the uses and abuses of numerical measures. When we think about numbers through the prism of the ethics of communication, two things become clear. First, the mere fact that a number is epistemically accurate is neither necessary nor sufficient for the communication of that number to be ethically justified. A number may be epistemically problematic, but permissible to communicate because all involved in the communicative transaction are (at least implicitly) willing to be deceived. Conversely, a number may be epistemically impeccable, but its use may still be manipulative. Second, the mere fact that using a number has positive consequences, regardless of its epistemic adequacy, does not show that it is ethically unproblematic. Thinking about numbers in terms of deceit and manipulation allows us to recognize that discussions of the limits of the numerical are no more, but also no less, than discussions of the messy ethical and political problems of treating each other well.

CHAPTER 7

Are Numbers Really as Bad as They Seem?

A Political Philosophy Perspective

Gabriele Badano

Many disciplines from the humanities and social sciences pay focused attention to the role of numbers in relation to political institutions. For example, historians have looked at how the rise of quantitative disciplines such as statistics and numerical tools like cost-benefit analysis has been inextricably linked to (and effectively driven by) the evolution of public administration (Desrosières 1998; Porter 1995). Anthropologists have analyzed the way in which practices including commensuration and the proliferation of numerical indicators constitute specific modes of power (Espeland and Stevens 1998; Merry 2016). Political scientists have investigated how numerical targets and other quantitative measures crucially contribute to governance in many political systems (Bevan and Hood 2006; Boswell 2018). This list could be much longer. However, there is at least one discipline that does not come easily to mind and whose inclusion would seem like a stretch—political philosophy.

If we pause for a moment to think about the fact that quantification is not a prominent subject of analysis for political philosophers, we will likely be puzzled by it. Here I am talking about what is normally referred to as "mainstream" or "normative" political philosophy. Dominating English-speaking academic debates in the field since the 1970s, this broad approach is interested in the evaluation of institutional arrangements from a philosophical perspective and in the provision of positive recommendations when applicable. Due to this normative focus, political philosophy seems highly relevant to discussions about the role of quantitative tools in political decision-making and possible comparisons with more qualitative approaches. Indeed, political philosophy seems very well placed to contribute, among other things, to the critical analysis of the ills created by too heavy a use of numbers in law- and policymaking and how best to balance qualitative and quantitative tools in order to foster important values institutions should pursue.

The overarching goal of this chapter is to finally make political philosophy part of the conversation. In order to do that, I concentrate on a debate that, although not explicitly focused on numbers, has interesting implications for the normative evaluation of quantitative tools in law- and policymaking. In tracing these implications, I will describe how the loudest voice from political philosophy turns out to be one that warns against the use of numbers in political decision-making. However, an important part of what I aim to achieve in this chapter is to argue that we should move beyond that voice and acknowledge that quantification might also play functions that are very important from a normative perspective, complicating the picture of its evaluation.

Section 7.1 focuses on the limits of precision as explored by the supporters of the influential "capability approach" to justice or, more specifically, to its evaluative space. It reconstructs two main arguments that capability theorists advance to support their conclusion that we should be wary of excessive precision and, therefore, too heavy a use of numbers in political decision-making. Section 7.2 digs somewhat deeper into political philosophy, demonstrating that, although less obviously connected to quantification, a prominent family of critiques of the capability approach is also relevant to the question of whether institutions should choose more qualitative over more quantitative tools to make decisions. These critiques, centered on the requirement that institutions should pursue justice in a publicizable way, can be developed so as to strike a far more positive note about numbers than the capability theorists' arguments do. Quantitative tools emerge as uniquely well suited to reduce complexity, potentially making justifications for laws and policies more manageable by the public. Finally, section 7.3 introduces a case study to reinforce the theoretical arguments advanced in the foregoing section. Specifically, it examines the National Institute for Health and Care Excellence (NICE), a British administrative body in charge of appraising drugs and other health technologies, to illustrate the idea that quantitative tools can foster public justification and even pose an obstacle to market forces in their dealings with institutions.

My argument is connected to several other contributions to this volume, especially, although not exclusively, those that are most philosophical. Like Greg Lusk (chapter 9 in this volume), I aim to show that, despite all the skepticism surrounding our discipline, academic work in philosophy can help further discussions about important social problems, including the ones concerning quantification. The connection to Stephen John (chapter 6) is even closer and extends to the substance of our analyses. A point that I wish to make is that political decision makers should at times use numerical measures to make an argument simpler in order to render it publicly available. This point has many similarities with John's notion of "spurious precision," which—as exemplified by the policy to encourage the consumption of five fruit and vegetables a day—consists in picking a single number while available evidence points to a broader range of values as the correct one. John explores the conditions under which it is ethical to resort to spuriously precise numbers in a way that is potentially very relevant to a research direction that I sketch but leave open in this chapter, focused on the ethics of using numbers for the purpose of simplification in the context of public justification.

7.1. CAPABILITIES, PRECISION, AND WORRIES ABOUT NUMBERS

As stressed in the introduction, political philosophy is somewhat exceptional in that the contrast between quantitative and qualitative tools to make political decisions has never become a topic of focused interest. This, however, does not mean that existing theories have no implication for the evaluation of quantitative tools in law- and policymaking. In particular, the limits of precision, which have obvious relevance for the role that numbers should play in decision-making, are an important theme in the capability approach to political philosophy. Given how prominent the capability approach is in the discipline, and how recurring this theme is, the loudest voice coming from political philosophy is therefore one of warning against an excessive use of numbers. But let us now take a step back and briefly reconstruct what the capability approach is.

The main idea behind the capability approach is that we should look at a specific evaluative space when discussing justice or, in other words, when working out what should be redistributed among citizens or otherwise guaranteed to them. The capability approach is distinctively focused on what is sometimes called "midfare," falling in between welfare economics' traditional focus on the utility generated by different states and the focus on the external resources—like money and access to education—that are needed for persons to enter those states (Cohen 1993, 18). According to capability theorists, the focus must be on the state of persons itself, keeping in mind what flourishing human beings do and are. Capability theorists generally stress that human flourishing is made up of many different elements, including good bodily health, the satisfaction of sexual desires, being imaginative and playful, having close friends, and being politically active. As a result, they also underline that the capabilities to do and be what is authentically valuable, which should be provided to citizens by political institutions, are plural.

The two most important capability theorists, Martha Nussbaum and Amartya Sen, are often at pains to point out that precision is frequently beyond our reach when we reflect about the ends of politics and make decisions about them. These authors echo a tradition of thought that reaches as far back as Aristotle, who famously said, while reflecting on the method of his *Nicomachean Ethics*, that we should be only as precise as our subject matter allows us to be; when it comes to political matters, we can draw conclusions only "sketchily and in outline" (Aristotle 2000, 1094b–95a). According to Nussbaum and Sen, being vaguely right in this area is definitely preferable to being precisely wrong, which puts into question the precision afforded by the use of quantitative measures in law- and policymaking.

We can reconstruct at least two arguments supporting Nussbaum's and Sen's claims that precision is often doomed to be misplaced and should therefore be avoided—an argument from ambiguity and one from incommensurability. Starting with ambiguity, Nussbaum maintains that philosophers as well as politicians and administrators are allowed to make generalizations in talking about elements constitutive of the good life and making political decisions aimed at fostering them. However, such generalizations must remain vague, leaving room for multiple specifications of each component of human flourishing. Not only individual societies but also individual groups in society and single individuals retain the authority to specify for themselves what it means, say, to build strong relationships with others or to be politically active. This is motivated by obvious liberal concerns with the oppression of minority groups. Also, Nussbaum appears to believe that the issue of specification of human flourishing simply admits of multiple answers; drawing again on the Aristotelian tradition, she suggests that the best way to tackle that issue is to let different groups in society tap into the rich intellectual resources provided by their specific religious and cultural traditions.[1]

The incommensurability argument does not target precision or, for that matter, the use of numbers as such. However, it does target a way of simplifying complex matters and bringing to the table a heavily numerical approach that is highly influential in academic disciplines like normative economics as well as in political decision-making, namely, commensuration. By commensuration I mean the identification of a single metric that will be used to evaluate states of affairs and choose between different courses of action. As I have already mentioned, the capability approach is committed to the idea that a flourishing human life is made up of plural components. Moreover, such components are thought to be qualitatively different from each other. Therefore, they are not reducible to one another and even less so to utility or any other generalist metric of value. Therefore, when it comes to evaluating a state of affairs or comparing alternative courses of action, institutions will have to be attentive to the provision of each basic capability in its own right, without any attempt at commensuration (Sen 1987, 61–65).

Both these arguments, and the one from incommensurability in particular, are closely connected to Nussbaum's and Sen's classic attacks on gross domestic product, especially when used to compare quality of life in different countries for the purposes of development policy, and on cost-benefit analysis as a tool for allocating scarce resources in global as well as domestic policy. To attach a monetary value to human well-being is thought to create a problem of commodification, making key elements of human flourishing alienable for a price. More generally, Nussbaum's and Sen's work often gives the impression that to resist precision (and numbers) in political decision-making is to resist the march of market forces in society. For example, Nussbaum enlists Karl Marx in her critical analysis of precision, accepting his claim that alienation comes not only from attaching monetary value to well-being but also from commensuration: "Assume the human being to be the human being and its relation to the world to be a human one, then you can exchange love only for love, trust for trust, etc."[2]

7.2. FROM PUBLIC JUSTIFICATION TO SIMPLIFICATIONS AND ON TO NUMBERS

Given that the use of numbers and, therefore, quantitative tools to make law and policy decisions involves precision, the capability approach's attack on precision clearly cautions against too great a use of numbers in political decision-making. The next step in my argument is to dig somewhat deeper into the intellectual resources provided by political philosophy, demonstrating that, although less obviously connected to numbers, an important objection to the capability approach also has significant implications for the role of quantitative tools in law- and policymaking. In tracing these implications, I aim to strike a more positive note about quantification than my argument has done so far. This is not to deny that Nussbaum's and Sen's arguments have merit. However, I believe that there is at least one more side to the story of quantitative tools in political decision-making than that of the alliance between numbers and market forces.

7.2.1. Taking the Steam out of Nussbaum's and Sen's Arguments

To give a little background, the objection to the capability approach that I aim to concentrate on comes from supporters of so-called resourcist approaches to justice's evaluative space, which I mentioned very briefly in section 7.1. Like capability theorists, resourcists reject the idea that, as a matter of justice, institutions should focus on the level of utility experienced by citizens. In contrast to capability theorists, however, they also disagree that institutions should try to work out, democratically or otherwise, a "thick" theory of well-being for their population that includes the essential components of human flourishing. Resourcists believe that institutions should abstain from setting the normative ends of human life and focus strictly on the provision of all-purpose means that can be useful to citizens with life plans oriented toward widely different ends. These all-purpose resources—which typically include civil and political liberties, access to health care and education, and income—make up shorter lists of goods than Nussbaum's and Sen's capabilities.

John Rawls is a very prominent resourcist thinker. When explaining why resources should be preferred over capabilities as the currency of justice, resourcists working in Rawls's tradition often make the point that institutions are not required simply to pursue justice, but to do so in a publicizable way. In other words, the goods populating justice's evaluative space must work as part of an authentically public criterion of justice. Why is this important? Thomas Pogge focuses on the fact that whenever institutions allocate funds to meet a citizen's needs, the public at large will bear the cost of this choice, both by paying the necessary taxes and by missing out on the benefits that could have accrued to them if funds had been allocated otherwise. According to him, widespread costs create an obligation on the part of institutions to be able to widely justify their choices (Pogge 2002, 208–9). An alternative and potentially complementary justification goes back to Rawls, who argues that public justification is necessary for reconciling the democratic idea that the members of the public hold ultimate authority over laws and policies with the fact that, in complex societies, they routinely have to live with laws and policies designed by others (Rawls 1993, 135–37).

Now, in Rawls's words, "simplifications" are necessary in order for any criterion of justice to become public (1999, 79).[3] To publicly justify the arrangements aimed at raising the prospects of the worst off in society, or society's institutional scheme more generally, is to choose arrangements that the public can broadly agree on. However, determining what are the components constitutive of human flourishing or, more broadly, what are the ends of life, is extremely controversial, as is the question of how to arbitrate the conflicts between such ends when all feasible courses of action can advance only some of them at the expense of others (Pogge 2002, 208–13). Therefore, according to resourcists, institutions cannot concern themselves with the high talk of ends and should rather settle for the provision of mere all-purpose means that virtually everyone can be expected to want for themselves whatever their life plans. Unlike capabilities, all-purpose resources have the potential to become publicly recognized objectives of government.

Thus far in this subsection, I have simply reconstructed a classic argument against the capability approach. This argument, however, is not normally brought to bear specifically on Nussbaum's and Sen's attacks on precision. I will now make this connection, suggesting that my argument allays many of the concerns animating Nussbaum's and Sen's objections to precision and, therefore, quantitative measures in political decision-making. Their claim that precision sweeps under the rug crucial ambiguities surrounding human flourishing loses its force if we accept that the job of institutions is simply to provide mundane resources such as money, access to education, and health care, leaving thicker accounts of well-being for individual citizens to figure out. Similarly, the capability theorists' worries about incommensurability are centered on the incommensurability of the different constitutive components of human flourishing, which, however, should not be the business of institutions.

In sum, on a resourcist account, the numerical tools that institutions are (at least in principle) supposed to use simply do not measure the thick version of well-being that is integral to Nussbaum's and Sen's ambiguity and incommensurability arguments against precision. Another way of putting this point is to say that Nussbaum and Sen do not really warn us against too wide a use of numbers in general; they only warn against too heavy a reliance on numbers provided that we accept their view of justice's evaluative space, namely that the aims of law- and policymaking (and, in turn, the typical constructs to be measured by the numerical tools employed in a political context) should essentially have to do with human flourishing.

At this point, some might object that agreement over political decisions and their justifications is simply too much to ask, casting doubt on the argument that I have outlined in this subsection and the way it rescues precision from the attacks of capability theorists. Rather than evaluate this objection, I will suggest that we could strike a positive note about numbers even if we decided to accept it. In the process of doing that, I will give a new spin to several keywords that have emerged during my reconstruction of the resourcist critique of the capability approach.

7.2.2. Redefining Public Justification

Let us then redefine public justification in a considerably less demanding way, as the ability on the part of law- and policymakers to place a justification for their choices "out there," in front of the public. Moreover, for any justification to be truly addressed to the members of the public, transparency in the disclosure of relevant information cannot be enough, but must be coupled with an effort to make the provided justification manageable by them. Among other things, manageability excludes scenarios in which the public is simply flooded with and left bewildered by a huge amount of information, for instance due to the complex nature of the factual side of the law or policy issue at stake. The commitment to manageability I am introducing echoes classic accounts of public justification. For example, in an analysis by Jeremy Waldron (1987) that is widely discussed as a powerful description of what public justification liberalism amounts to, the mark of a well-ordered liberal society is that the principles governing its institutions are "available for public apprehension" (146) or "capable of explaining [themselves] at the tribunal of each person's understanding" (149).

Repurposing another Rawlsian concept, there is room to argue that public justification as I have redefined it often requires simplifying devices that allow decision makers to provide the public with the skeleton of the reasoning behind a law or a policy. When complex issues are on the table, such devices are the only hope to produce justifications that members of the general public can reasonably be expected to manage, as opposed to perceive as spoken largely over their heads. Also, reducing justifications to their bare bones is often important for starting democratic deliberation across society. This is where we get back to the topic of quantification and quantitative measures, which by their nature reduce their objects to a few key dimensions while excluding others, and which are therefore very well placed to serve as the simplifying devices that public justification needs.

Now, a reader might ask, is this simplifying power really unique to numbers? The use of words in putting together a justification for a political decision already involves a degree of abstraction from the problem in question and the proposed solution as actually experienced, in its full complexity, by flesh-and-bone individuals. I do not mean to deny this. At the same time, however, it is hard to find any scholar of quantification who fails to stress how numbers are characterized by a uniquely strong tendency to leave behind some of the features of their intended objects of measurement. This, I believe, makes intuitive sense: if qualitative descriptions of social phenomena necessarily involve abstraction, it stands to reason that a particularly strong form of abstraction is involved in their quantification.

To quote but a few examples from the literature, Theodore Porter makes some points that go in this general direction when discussing the measurement of social as well as natural objects. Among many other cases, he describes how, before thermometers were perfected during the eighteenth century, the concept of temperature referred to a rich variety of features of the atmosphere as perceived by people. "To quantify was necessarily to ignore a rich array of meanings and connotations," he writes, given that in general "quantification depends on the art of forgetting" (Porter 1994, 396). On her part, Sally Engle Merry focuses on the numerical indicators that are employed to monitor the implementation of human rights worldwide. Her very definition of numerical indicators depicts them as meant to turn complex data into a simple number (Merry 2011, S86–S87); for her, one of the overarching functions of numerical representation in public policy is to simplify complexity (Rottenburg and Merry 2015, 7). In fact, according to Merry, numbers tend to flatten individual differences, gloss over context, and lose information more generally.

Most of the authors discussing the link between quantification and simplification do it in explicitly critical terms. This holds true for Merry and even more so for other scholars. Drawing on Merry's work, Jerry Muller (2018) complains about the tendency of metrics to restrict attention to a few aspects of their target social phenomenon, which often are the simplest and easiest ones to measure (23–27). Lorenzo Fioramonti (2014a) compares those who try to understand the economy through statistics to the blind characters in an Indian fable who try to learn what an elephant is by each touching one single part of it. The message that Fioramonti aims to convey is that gross domestic product and other available statistics focus only on small, isolated aspects of the economy, like the man who touches the trunk of the elephant and thinks he is faced with a snake, or the one who touches its leg and thinks the elephant is a tree (16–17). For Cris Shore and Susan Wright (2015), who look mostly at audits, "tunnel vision" is one of the main negative consequences of quantification (430–31).

Through the link I have drawn between public justification, manageability, and the need for simplifying devices and numbers, my aim has been to push back, at least to some extent, against this overwhelmingly negative assessment of the simplifying powers of numbers in political decision-making. Quantification is very well placed to help decision makers meet the need for public justification. This fact demonstrates that the tunnel vision induced by numbers can play very important roles and is not to be discarded lightly. Obviously, this point is not enough to cancel out the possible dangers associated with numbers that I have hinted at in the previous two paragraphs. More than anything else, this point highlights

the need for further research—research into the ethics of using quantification to reach the level of simplicity required for justifications to be truly
manageable by the public.

Although very important, this line of research is extremely complicated
and therefore falls beyond the scope of this chapter. In concluding this section, I have room for only a few preliminary remarks about what it might
mean to use the simplifying powers of numbers ethically. An ethical theory
of quantification in this area should probably start with a few clear statements about how numbers must *not* be used. Decision makers should not
make a justification for a law or policy manageable by the public by simply
zooming in on its simplest aspects, or the ones easiest to measure, regardless of whether they are central to their argument for that law or policy.
Even more importantly, the introduction of numbers into a justification offered to the public should not be motivated by merely rhetorical purposes.
Numbers have a well-known allure (Seife 2010), but using them to simply
move the public to accept a proposal, without aiming to add anything to
their understanding, is inconsistent with public justification properly understood. Decision makers should use quantification to narrow the focus of
their justifications for laws or policies to the few aspects that represent the
innermost core of their argument. Their goal should be to reduce complexity to the extent necessary in order for the resulting justification to become
manageable to the majority of the general public in the sense of the term
explored in this subsection. To reiterate, this is but a preliminary description of what an ethical use of numbers in public justification amounts to,
subject to revision and expansion in future investigations.[4]

7.3. THE CASE OF NICE

Section 7.2 suggested that when it comes to normatively appraising the
role that quantification should play in political decision-making, things
are more complicated than the capability theorists' arguments against
precision appear to indicate. The task of this section is to reinforce the
theoretical conclusions reached in section 7.2 by looking at the use of numerical measures made by a real-world administrative body, the one in
charge of evaluating pharmaceutical and other health technologies for use
in the National Health Service (NHS) in England and Wales.

7.3.1. QALYs, NICE's Cost-Effectiveness Threshold, and Manageability

The administrative body I will discuss is the National Institute for Health
and Care Excellence (NICE). Created in 1999, NICE provides guidance

to the NHS in several areas, but here I will focus exclusively on its health technology appraisal (HTA) process. Through this process, NICE recommends whether or not local NHS commissioners should purchase the health technologies that have been referred to it for appraisal. And if NICE recommends a technology, commissioners are legally bound to fund it under the NHS. Given the crucial role played by its HTA process within the NHS, NICE is often in the spotlight, both in civil society and among academics. In part because NICE itself has a clear sense that its methods are grounded in moral or otherwise philosophical assumptions (National Institute for Health and Care Excellence 2008), it has also been widely discussed (and criticized) from a philosophical point of view (see, e.g., Harris 2005a, 2005b). Political philosophers have been part of this conversation, bringing theories of the so-called pattern of distributive justice and of fairness in deliberative democratic decision-making to bear on the work of NICE and other healthcare resource allocation bodies (Daniels 2008). However, no one has so far brought together NICE and topics like precision and simplicity from the debate over justice's evaluative space.

When evaluating a drug or some other health technology, NICE calculates the added cost to the NHS of producing an additional unit of gain in health-related quality of life through that technology compared to the technology that the NHS currently uses for the same indication. In technical language, this is the appraised technology's "incremental cost-effectiveness ratio." NICE's measure of health-related quality of life is the quality-adjusted life year (QALY), which integrates the increases in life expectancy and the gains in quality of life brought by healthcare treatments. In so doing, the QALY reduces to a single numerical scale a huge variety of benefits, which are as diverse as those produced by life-saving surgery, life-extending drugs for terminal cancer, medications that slow down dementia, palliative treatments, surgery capable of restoring mobility, and treatments for depression. The QALY represents a heroic attempt at commensurating very different objects. Consequently, it provides a first excellent example of a numerical tool displaying the traits of a simplifying device.

The centerpiece of NICE's HTA method is the comparison between the incremental cost-effectiveness ratio of the technology under appraisal and NICE's soft incremental cost-effectiveness threshold, ranging from £20,000 to £30,000 per QALY. NICE's threshold can be read as another example of a numerical measure simplifying an enormously complicated object. Indeed, it can be read as a measure of (or at least a first approximation of) good value for NHS money. If a technology costs less than £20,000 per QALY, NICE is extremely unlikely to reject it. NICE can recommend a technology whose incremental cost-effectiveness ratio falls in the £20,000–£30,000/QALY range, provided that support for that technology is offered by any item on a list of additional considerations—severity of disease; benefits delivered at the end of life, on which there is a premium; disadvantaged groups, to whom special attention is paid; benefits accrued to children, which are given extra weight; and several others. NICE can approve technologies above the £30,000/QALY mark if such considerations provide an exceptionally strong support to them (National Institute for Health and Care Excellence 2013).[5]

A paper that I coauthored with Stephen John and Trenholme Junghans (Badano et al. 2017) explores the process through which NICE arrived at the £20,000–£30,000/QALY figure for their cost-effectiveness threshold. The justification that health economists normally provide for NICE's use of a threshold (and that NICE explicitly endorses) is that the threshold is a measure connected to the opportunity costs of newly funded technologies against the background of the NHS's finite budget. Specifically, it is meant to measure the cost-per-QALY level beyond which a newly funded technology would displace greater QALYs across the NHS through disinvestment than it would produce. Going back to a comment I made in the previous paragraph, the threshold is supposed to measure the level beyond which the technology under appraisal counts as insufficient value for money in terms of net QALY output. To serve this function, the threshold should be set at the level of the cost of a QALY produced through the least cost-effective treatment currently funded under the NHS (McCabe et al. 2008).

However, John, Junghans, and I found that when, in 2001, NICE set its threshold at £20,000–£30,000/QALY, it was nowhere near having the necessary empirical information to identify how much it cost to produce a QALY through the least cost-effective treatments that the NHS was providing. In fact, NICE did not have a solid evidence base to set its threshold until much later, around 2013. What NICE did to identify the £20,000–£30,000/QALY figure was to look back at the HTA recommendations it had issued case by case until 2001, based on its decision makers' best insights. The figure derives from the simple fact that NICE found that while rejections had been very rare below £20,000/QALY, the likelihood of a negative recommendation increased above it, and a positive recommendation had been extremely unlikely beyond £30,000/QALY (Badano et al. 2017, 155–58).

From NICE's stated health-economic perspective, the choice of the £20,000–£30,000/QALY figure seems to make little sense. Still, we argue that once we switch to a political philosophical perspective, there are things to be said in favor of NICE's choice to nonetheless pin down explicit numbers for its threshold, at least until the relevant evidence emerged in 2013. For the sake of my argument in this chapter, those things link back to points made in section 7.2.

The choice to place the £20,000–£30,000/QALY figure at the very center of NICE's HTA process made it much easier for NICE to explain to the public how its decisions were made, and thus where this or that recommendation against new technologies came from (Badano et al. 2017, 159–63). This fact increased both public transparency and what this chapter has called "manageability by the public," in turn contributing to NICE's ability to publicly justify its decisions in the less demanding sense of the term outlined in section 7.2.2.

To better see how manageability was helped by NICE's choice, just imagine an alternative scenario in which NICE abandoned its plan to place at the core of the HTA process its best estimate of any numerical measure of good value. We can imagine that NICE made this choice in 2001, either for lack of a solid evidence base or because they could not stand the simplifications involved in both commensurating hugely different benefits on the QALY scale and giving a place of honor to any single measure of value for money. In that scenario, NICE's recommendations for or against any new technology would become largely a matter of judgment by those in charge of the decision, to be exercised in identifying all the relevant considerations for the case under appraisal, in solving their conflicts, in working out how the context of the specific decision changes relevant considerations and how to balance them, and the like.

It is extremely difficult, if not impossible, to imagine how decision makers could transparently explain their judgment calls to others.[6] Even if we assumed that they could, such calls would deal with a complex plurality of considerations and be highly sensitive to context, creating the problem of flooding the public with overly complex information which I underlined in section 7.2.2. In other words, by giving up the numerical centerpiece of its decision-making method, NICE would have given up the solid backbone it has traditionally been able to go back to in order to explain its choices to the public in a way that is simple enough to be largely manageable by them.

Moreover, John, Junghans, and I suggest that the creation of a numerical threshold and the publicization of decisions made on its basis have served as a catalyst for a broad democratic debate in civil society over the allocation of NHS resources—a debate whose scale is arguably unparalleled across the world and that is credited by some for having convinced large sectors of the British public of hard truths, including the need to ration healthcare resources (Badano et al. 2017, 159–63). Whatever we might want to say in favor of NICE's choice in 2001 to pin down an explicit figure for its threshold while waiting for in-depth empirical research, things changed in 2013 when NICE decided to stick to its £20,000–£30,000/QALY figure even though the much-awaited evidence had just made it clear that the threshold should actually be lowered to around £13,000 per QALY. From that point on, the choice to use a £20,000–£30,000/QALY threshold has created a brand-new contradiction between the decisions made by NICE about new technologies and the justifications NICE has kept issuing for them—justifications that, still describing the threshold as meant to identify the cases in which new technologies would displace more QALYs than they would create, cannot consistently incorporate a threshold set as high as £20,000–30,000/QALY. As politically hard as it might be, public justification appears to require that NICE revise its threshold and make the difficult choice of lowering it.

7.3.2. QALYs, RCTs, and Numbers as Obstacles to Market Forces

Let us now turn to a more recent development in NICE's history, which is especially relevant to the alleged alliance between numbers and market forces that emerged during my discussion of capability theorists in section 7.1. In 2016, NICE took over and completely restructured the Cancer Drugs Fund (CDF). The CDF is a ring-fenced fund with an initial annual budget of £340 million, to be used by NICE to temporarily fund cancer drugs that would have otherwise been rejected, therefore enabling the collection of new evidence during the process. Now that the CDF is in place, NICE can recommend a cancer drug for usage under the NHS, albeit under special arrangements, even if NICE decision makers believe that its cost-effectiveness is still very uncertain. These arrangements require that over the following two years, the company producing the drug will have to collect further evidence aimed at reducing the existing uncertainty. At the end of this period, NICE decision makers will reconvene to decide whether the new evidence is sufficient to warrant confidence in the cost-effectiveness of the drug under appraisal and, therefore, whether it should be discontinued or recommended for routine commissioning under the NHS (NHS England Cancer Drugs Fund Team 2016).

This reform to NICE's HTA process is interesting in many respects. However, I wish to focus on the data collection arrangements between NICE and the producers in charge of dispelling the uncertainty surrounding their drugs. These arrangements break with NICE's traditional method in giving priority to observational studies over randomized controlled trials (RCTs) when it comes to determining the QALYs produced by drugs and, in turn, whether or not such drugs meet NICE's £20,000–£30,000/QALY threshold. RCTs, the cornerstone of evidence-based medicine, measure statistical associations, and thereby provide a highly quantitative method for investigating effectiveness. Given that pharmaceutical producers have only two years to collect evidence while their drugs are funded under the CDF, however, NICE states that "careful consideration" must be given to any plan to start new RCTs (National Institute for Health and Care Excellence 2016, 9). Therefore, observational data, obtained through NHS registries that systematically collect information about cancer patients and their responses to treatment, may well be the only evidence submitted to NICE at the end of the two-year period.

NICE's choices about data collection arrangements did not happen in a vacuum. They were inspired by an increasingly influential model for the collection of medical evidence, which is critical of the rigidity of RCTs and advocates a switch to flexible "adaptive" pathways to evidence gathering and "real-world" observational studies. Many commentators note, however, that RCTs are rigid for a reason: the tight constraints imposed on evidence collection are necessary for minimizing manipulation by researchers with vested interests in the outputs of their studies, like the producers in charge of collecting observational data under the CDF—or, for that matter, the great many clinical scientists worldwide who are funded by the pharmaceutical industry (Grieve et al. 2016). Also, observational data are known to be worse than RCTs at countering biases toward positive results, increasing the likelihood that researchers will conclude that a drug is beneficial even if it is ineffective or even harmful. In turn, this increases the likelihood of NICE mandating that the NHS pay pharma for drugs on exit from the CDF that are either useless or outright harmful for patients (Davis et al. 2016; see also Howick 2011, 39–62).

A powerful distillation of these arguments is offered by some critics of this emerging model, who point out that accepting adaptive pathways effectively amounts to "adapting to industry" (Davis et al. 2016). Obviously, this point is controversial, and the supporters of adaptive and real-world approaches dispute the objections that are being raised against them.[7] Still, their critics surely manage to make an appealing counterpoint to the view I have extrapolated from the capability theorists' critique of precision, namely, that the march of numbers in political decision-making is one and the same as the march of market forces in society.

A common complaint about algorithmic or simply numerical approaches to political decisions is that they are often too inflexible. Still, we can now see how this rigidity might be an important asset in resisting market forces.

In certain contexts, a heavily quantitative and therefore fairly rigid approach, which will capture nothing more than a simplified version of quality of life or any other value it is supposed to serve, appears to provide the only way to constrain industry in its dealings with policymakers, by forcing them to actually show that they can deliver at least some kind of value to the public. In contrast, due to industry's economic and lobbying powers, a flexible and more qualitative approach risks leading to a scenario in which industry systematically has its way, to the detriment of other stakeholders like patients, taxpayers, and citizens at large. As far as NICE is concerned, a "numerical connection" linking attention to RCTs to the use of QALYs and finally to NICE's threshold seems able to do good work and is, in fact, under attack by the pharmaceutical industry. This is despite the fact that, as I mentioned at the end of section 7.3.1, the threshold is already too generous with the new drugs that producers bring to the market.

7.4. CONCLUSION

Political philosophy seems highly relevant to existing debates over the role of numbers in political decision-making. However, quantification has not been the subject of focused attention within the discipline. The overarching goal of this chapter is to take a first step in the direction of filling this gap. Given the close link between numbers and precision, the critique of precision that recurs in the work of prominent capability theorists makes the loudest voice from political philosophy one of warning against quantification. However, by drawing on objections to the capability approach that stress the importance of public justification and, in turn, by simplifying devices in political decision-making, I argue that quantitative tools should not be dismissed quickly. The case of NICE has reinforced the impression that such tools can indeed perform functions of great importance within political institutions, strengthening the normative conclusion that in some circumstances they might well be the best choice for making law- and policy decisions.

If looked at from the perspective of political philosophy, the argument of this chapter is surprising. The main framework I employ to identify the positive functions that numbers can play is that of public justification, which owes much to the classic work of Immanuel Kant. As evident with authors like Rawls, many descriptions of public justification retain a distinctive connection to Kant's first formulation of the categorical imperative, requiring that agents check that their maxims can be willed as universal laws in the sense that they must be acceptable from all perspectives (Vallier 2018). Broadly Kantian approaches to political philosophy, however, are normally pitted against, not classed as, number-heavy theories, which are most obviously exemplified by utilitarianism. The surprising bit of my argument is that there is a strong link between a Kant-inspired commitment to public justification and numerical tools in political decision-making, which, I suggest, are extremely well suited to simplify arguments to the extent necessary to make them public. This is not to say, of course, that the use of numerical tools is without dangers, from a Kantian or other perspective. I have highlighted the need for an ethical theory of how such tools are to be used—an urgent topic that deserves much focused attention.

NOTES

1. Nussbaum (1992), 224–25. Out of respect for real-world democratic processes, Sen (2004, 333) goes a step further than Nussbaum, rejecting also the idea that scholars should try to draft a list of basic capabilities.

2. Marx and Engels (1978), quoted by Nussbaum (1992), 231. For another argument that the role of numbers in politics is to entrench the power of markets, see Fioramonti (2014a), 1–9.

3. Pogge (2002, e.g., at 208) sometimes speaks in terms of the "specificity" required by a public criterion of justice, bringing up another desideratum that is connected to the use of quantitative tools.

4. Interesting resources to continue this line of research could be provided by Stephen John and Anna Alexandrova, both writing in this volume, looking respectively at the ethics of spurious precision and at how to evaluate the effects on important values of political attempts to redefine them so as to build in quantifiability, comparability and so forth.

5. For a focused discussion of the considerations to be taken into account in addition to cost-effectiveness, see Rawlins et al. (2010).

6. For a classic account of the way in which judgment interferes with transparency, see Richardson (1990).

7. For example, see Guido Rasi and Hans-Georg Eichler's letter (2016), written in response to the critics of adaptive pathways.

The Uses of the Numerical for Qualitative Ends

CHAPTER 8

When Well-Being Becomes a Number

Anna Alexandrova and Ramandeep Singh

Quantifying well-being is an old ambition. Proposals for how to measure happiness were made in the Age of Enlightenment by utilitarian philosophers, throughout the nineteenth century by classical economists, and in the twentieth century by social scientists of many stripes. Nevertheless, these initiatives remained the province of the quirky theoretician and utopian, while well-being and happiness carried on being principally subjects of art, literature, philosophy, religion, and personal reflection, where measurement was not the point. This began to change in the late twentieth century.

The comfort with quantification of what is perhaps the ultimate personal phenomenon was an outcome of several trends (see Angner 2011; Davis 2015; Gere 2017). First of all, it is a culmination of decades of academic work at universities and commercial laboratories refining measurements of psychological traits, emotions, and attitudes. Questionnaires and psychometric scales of positive states such as happiness and satisfaction proliferated and a new identity of "positive psychologist," or "well-being scientist," emerged. Central to this identity is the development and validation of such scales and their use in experimental and statistical studies of the determinants of well-being. The second contributing trend was the self-help movement that aligned itself with experimental psychology rather than the earlier humanistic tradition of psychotherapy and psychoanalysis. This movement encouraged production of popular materials such as books, training programs, and, lately, digital apps, which eventually made their way into management, human resources, and life coaching. Finally, the last decades of the twentieth century saw the rise to prominence of critiques of orthodox economics and growing demands that evidence-based policy be responsive to more than just growth of gross domestic product (GDP), consumption, and income.

The late 1990s and the early 2000s saw high-profile conferences and publications in which eminent US economists and psychologists such as Daniel Kahneman, Ed Diener, and Martin Seligman touted the optimistic new science of the good life.[1] In 2009, three famous economists, Joseph Stiglitz, Amartya Sen, and Jean-Paul Fitoussi, produced a report commissioned by then French president Nicolas Sarkozy outlining the importance of well-being in national accounting (Stiglitz et al. 2009). Guidelines for measurement were endorsed internationally and intersectorally by governments and nongovernmental organizations (NGOs), reaching even traditional economic tools of cost-benefit analysis (Fujiwara and Campbell 2011; Organisation for Economic Co-operation and Development 2013). The arguments used against GDP and in favor of richer measures were often old, resurrected, for example, from Robert F. Kennedy's 1968 speech that indicators such as gross national product measure everything "except that which makes life worthwhile." By 2012 such sentiments no longer sounded utopian, and the UK economist Richard Layard radiated confidence in *The Guardian*: "If you go back 30 or 40 years, people said you couldn't measure depression. But eventually the measurement of depression became uncontroversial. I think the same will happen with happiness" (Rustin 2012).

Today there is a critical mass of consensus that well-being is quantifiable, or at least that it can and should be represented quantitatively. Our focus here is the right attitude to this consensus. Should well-being quantification be exposed, opposed, and discouraged as distorting the true nature of this complex phenomenon and as giving in to the late capitalist dream of the numerical self? Or should it be tolerated and even celebrated as a scientific achievement centuries in the making? The answer, perhaps predictably, is neither. The triumphalist narratives miss the mark because the controversies about constructing a scale of so slippery a phenomenon never got resolved, only forgotten. Categorical disenchantment with well-being quantification is problematic too: while it is tempting to point to the mismatch between well-being properly understood in the light of some philosophical theory and the current measures, such criticism commits a category mistake. Well-being "properly understood" is not the target of these measures. Instead, their proponents redefine well-being in a way that builds quantifiability into the very concept. They sacrifice theoretical validity for the sake of making well-being a viable object of public debate.

The better question is whether this move of construing well-being as a quantitative phenomenon for pragmatic reasons is defensible *all things considered*. The issue as we see it rides on whether it is appropriate to trade off theoretical validity against effectiveness in public debate, by using a quantitative measure for a qualitative phenomenon. Because this conflict was articulated recently by one of the architects of British well-being measurement, politician Oliver Letwin, we call it "Letwin's dilemma." Each type of well-being quantification makes a trade-off between theoretical and practical demands, and different trade-offs are justifiable in different contexts. We therefore urge that quantification of well-being should be neither cheered nor denounced as a whole, but instead evaluated on a case-by-case basis. As an illustration of our strategy, we discuss two such cases. In the first one—the measure of UK national well-being by the Office for National Statistics (ONS)—well-being is quantified by a rich table of indicators which strikes, in our view, a defensible balance between validity and practicality. In the second—*The Origins of Happiness*, a report by economists at the London School of Economics—quantification runs amok, making too great a sacrifice of richness and complexity for the sake of political goals that are themselves questionable.

8.1. DIVERSITY OF QUANTIFICATIONS: A SURVEY

What does it mean to quantify well-being? The first step is to adopt a definition of well-being; the second is to put forward a scale that captures variations in well-being according to this definition. There are many such definitions available and several possible scales for each, so options multiply quickly. Table 8.1 summarizes what we see as the main traditions.[2] Each row presents a different bundle of measures that correspond to different answers to the initial question "What is well-being?" Very roughly, the first three rows come from the psychological sciences. While they represent distinct philosophical traditions, all elicit self-reports of well-being with questionnaires. Psychologists in the first row identify well-being with felt experiences of positive and negative emotions, or happiness, and they trace their intellectual roots to hedonism. They favor experiential measures of happiness that ask whether respondents feel a given emotion such as sadness or joy. These ratings are then aggregated into a "hedonic profile" that represents a time slice of the individual respondent. Those in the second row see well-being instead as an individual's judgment about her life as a whole, or her life satisfaction, and hence adopt short evaluation questionnaires that invite subjects to agree or disagree with statements such as "All things considered my life is going well." Their philosophical heritage is probably closest to subjectivism—the theory that grounds well-being in the fulfillment of the individual's priorities. Finally, the advocates of flourishing, in the third row, trace their roots to Aristotle (and classical eudaimonism more broadly) as well as to twentieth-century humanistic psychology: the good life is a life of maximal functioning and

TABLE 8.1. Definitions and measures of well-being

Definition	Measure
Happiness	Experience sampling; U-Index; Positive and Negative Affect Schedule; SPANE; Subjective Happiness Scale; Affect Intensity Measure
Life satisfaction	Satisfaction with Life Scale; Cantril ladder; domain satisfaction
Flourishing	PERMA; Psychological General Well-Being Index; Flourishing Scale; Warwick and Edinburgh Mental Well-Being Scale
Preference satisfaction	GDP; GNP; household income and consumption; stated satisfaction surveys
Quality of life	Human Development Index (capabilities); UK Office of National Statistics Measure of National Well-Being; Legatum Prosperity Index; Social Progress Index; OECD Better Life Index; Nottingham Health Profile; Sickness Impact Profile; World Health Organization Quality of Life; Health-Related Quality of Life

actualization of personal potential. Although operationalizing this theory is undoubtedly hard, psychologists typically articulate several "virtues," such as autonomy, connectedness, and sense of purpose, and ask respondents to answer questionnaires corresponding to each.

The definition of well-being as "preference satisfaction" in the fourth row comes from economics. The economic tradition of welfare measurement builds upon the view that well-being consists in satisfaction of the individual's preferences as expressed in their choices. Adding the assumptions that money is a measure of the individual's ability to satisfy their preferences, and that individuals make choices rationally, income (or consumption) becomes a proxy for well-being.

"Quality of life," in the last row, includes several different traditions reflecting different understandings of this concept. First, we have the "social indicators" tradition in sociology from the 1970s, which sought to enrich the statistics collected by governments and NGOs beyond the most basic ones. Second is development economics with its capabilities approach (and various related proposals), which sets out to capture sets of goods that matter for the progress of poor countries beyond mere economic growth. Finally, quality-of-life measures focusing specifically on health are common in medical and public health research. In all these cases, a measure is basically a collection of indicators all thought relevant to quality of life (including sometimes economic and subjective indicators) plus a rule about how to aggregate these indicators into a single number, if necessary.

To take stock, the table shows distinct traditions of defining and of quantifying well-being. Crucially, these traditions do not represent different ways of measuring the same phenomenon. Rather they conceptualize well-being differently. They disagree as to what phenomenon we are talking about when we talk about well-being. Emotional states are one thing, broader quality of life quite another, and there is no sense in claiming that a measure of a person's emotional state is superior, as an indicator, to a measure of their quality of life. Any choice between the rows of our table will be made on the grounds that one of these phenomena is considered the right focus of well-being research. The next two sections of this chapter report how one of these rows—namely life satisfaction—is in the process of becoming predominant.

8.2. IN SEARCH OF THE RIGHT QUANTITY

From the perspective of advocates of subjective well-being (that would be the first three rows), the villain was and remains the income and consumption measures of orthodox economics. Economists call these measures "indicators of well-being," and they come with a straightforward and well-developed story about quantification.[3] But the public face of the science of well-being is typically associated with a rejection, or at least an attempt to marginalize, the economic definition. "Beyond Money: Toward an Economy of Well-Being" was the title of a seminal 2004 article by US psychologists Ed Diener and Martin Seligman, wherein the new field of research was explicitly positioned in opposition to economic measures, whose problematic assumptions were also exposed.

The standard criticism seizes upon the "Easterlin paradox," named, ironically, after the economist Richard Easterlin, who first articulated it in the 1970s. He juxtaposed two facts: at any given time and within any country, income predicts self-reported happiness, but over time, as income increases, happiness does not rise correspondingly. To resolve this tension, Easterlin hypothesized that beyond a certain minimum, people's judgment of their well-being is indexed to their income relative to that of others—not in absolute terms (relative to their previous income level). Therefore income will fail to track subjective well-being over time. In the 1990s and early 2000s, the Easterlin paradox acted as a powerful motivator for research on the relationship between objective circumstances and life evaluation, which in turn helped establish the basics of well-being quantification. Now it is far less clear whether the paradox even exists, as new research does not find evidence for the second supposed fact—increase in absolute income does after all seem to predict increases in subjective well-being over time and hence it is unclear that money and happiness come apart as Easterlin (1974) claimed they do. But these findings have not managed to dampen the interest in self-reported happiness. Even if, on average, absolute income and subjective well-being rise and fall together, there are still striking cases of divergence, for example the steady growth of GDP coupled with a steady fall in life satisfaction in Egypt and Tunisia during the Arab Spring.[4] So there is still room for attending to subjective well-being as independent of income.

However, to make subjective well-being a genuine alternative for policymaking and evaluation, its advocates needed to turn it into a quantity just as manageable as the traditional indicators. Otherwise, in response to the passionate calls to make policy accountable to people's priorities, the economists could easily retort: "Well-being is all nice and good, but how will we plug it into budgets, spreadsheets, and cost-benefit analyses?" This demand for quantifiability undoubtedly stems from presuppositions about the nature of proper scientific evidence, presuppositions that can be easily challenged on philosophical grounds. But, as we shall see, there are political grounds for it too—considerations of democracy.

As table 8.1 shows, however, there is no single way of quantifying even *subjective* well-being, let alone well-being under other conceptions. If you are a hedonist, quantification of happiness will take the form of turning momentary reports of emotional state into a single rating, and that single rating into a dot on a curve that represents a subject's emotional state over time. Such "experience sampling" takes time and resources, and despite its faithfulness to classical Benthamite utilitarianism and some attempts to implement it on a large scale, even psychologists who are sympathetic to this philosophy agree that it is not practical. Although it continues to be used in research, it is not part of official statistics of the sort that well-being enthusiasts call for.[5] By and large, such statistics are based on questionnaires, especially those that Easterlin himself used—the life-satisfaction questionnaire.

8.3. LIFE SATISFACTION AS A QUANTITY OF WELL-BEING

Life-satisfaction questionnaires quantify well-being by asking respondents to agree or disagree (plus strongly agree, strongly disagree, or neither, and so on) with statements such as: "In most ways my life is close to the ideal" or "I am satisfied with my life."[6] When large samples of people take these questionnaires, it generates a lot of data finely differentiated by the degree to which people endorse a given statement. If these questions exhibit the right psychometric properties of reliability and predictability among the right populations of subjects, they are typically declared valid. It thus becomes possible to talk about the differences in life satisfaction between different groups (Finns are happier than Russians) and about life satisfaction coefficients (living in a safe community raises life satisfaction by that much). This is a major step toward making subjective well-being a quantity.

That life-satisfaction questionnaires can jump through these hurdles of psychometrics does not make them immune from criticism. Of these there are roughly three types. The first comes from the point of view of ethics and axiology. What does life satisfaction have to do with well-being? Philosopher Daniel Haybron (2008) argues that to be satisfied with life is by and large to endorse certain values such as gratitude, modesty, and determination. The extent to which we are satisfied with our lives is reflected in how thankful we think we should be for what we have or whether we think we should not rest on our laurels. Life satisfaction reflects "one's stance towards one's life" (89). But the stance we adopt toward our life on reflection is one thing, he maintains, and how we actually feel—our emotions in daily life—is another. A great deal of misery in daily life is compatible with high life satisfaction, which makes it implausible as a reflection of a person's well-being.

A second worry comes from the point of view of measurement theory. Ratings on a Likert scale strictly speaking justify only an ordering: if I rated myself as strongly agreeing with a statement about my life satisfaction and you rated yourself as agreeing only somewhat, then (assuming our ratings are comparable—more on that shortly) it is permissible to claim that I have a higher satisfaction with life than you. But that's all an ordering justifies. The conventions of metrology do not permit the more specific claim that I am more satisfied than you by x points; nor is it permissible to average our ratings in order to make population-level comparisons. To use technical language, ordinal scales should not be treated as cardinal ones, but it is hard to see how the strand of well-being science that is intent on providing an alternative to economic indicators can make do only with ordinal comparisons.[7]

Finally there are worries from the point of view of psychology. What sort of judgment is a judgment of life satisfaction? On the face of it, it requires an aggregation of a great deal of information on the part of the respondents: here are all the things in life I value, here's how well I think I am faring on each of those values, here's how all these assessments tote up when I think about my life as a whole. But do people actually perform such complex evaluations? There is a long-standing concern with the alleged fickleness of life-satisfaction judgments: apparently finding a coin, or seeing a person in a wheelchair, or being reminded of the weather, can drastically change a person's evaluation of their well-being. Such effects suggest alternative explanations for how these judgments are formed (perhaps they are made on the spot and are deeply susceptible to mood) and makes their replicability dubious (Schwarz and Strack 1991, 1999). And here we have not even broached another psychological controversy—whether life-satisfaction ratings are comparable between individuals or across cultures.

None of the three angles of attack have stopped life satisfaction from becoming the most popular measure of well-being today. Its advocates dismiss the philosophical worries on the grounds that ratings of life satisfaction correlate decently with longer and richer questionnaires and that respondents' own judgments about how to evaluate their lives should be respected (Cheung and Lucas 2014; Diener et al. 2013). The worries of measurement theorists are harder to dismiss but scientists disagree on their severity, with Ferrer-i-Carbonell and Frijters (2004) claiming that treating ordinal data as cardinal "does not generally bias the results obtained," while Schroeder and Yitzhaki (2017) argue to the contrary.

It is the psychological objections that scientists have taken most seriously and to which they have found relatively convincing replies. New experiments reveal that judgments of life satisfaction are actually quite robust. The finding about the weather/coin/wheelchair effect on life-satisfaction reports has not been replicated (Lucas 2013). The context in which people are asked to judge their life satisfaction—what they are thinking and experiencing at the moment and in what circumstances—clearly affects this judgment. But whether these context effects make these measures unusable and uninformative is far less clear.[8] There is also evidence of the interpersonal and cross-cultural comparability of life satisfaction (Diener and Suh 2003).

To take stock: life-satisfaction questionnaires are controversial but hard to abandon. While the theoretical case against them is strong, they are also widely available and generate interesting data. For practical reasons, none of the philosophical and methodological objections raised against them has been able to dislodge them from their dominant status in the science of well-being. In the OECD (2013) report with official guidelines for collecting well-being statistics, life satisfaction has a central place. Is this consensus a case of dangerous ignorance of facts for the sake of expediency?

8.4. LETWIN'S DILEMMA

So far, we have seen two facts that pull in different directions: on the one hand, there is a huge diversity in well-being concepts and measures, with no decisive theoretical reason to settle for one rather than another; on the other hand, life satisfaction is emerging as a winner on grounds that are often pragmatic (this will become even clearer in section 8.6). In this section, we complete the story of how this tension can be resolved by introducing the figure of the "political sponsor." While our example of such a sponsor is from recent British history, the tropes this figure uses illuminate the predicament more generally. The sponsor sidesteps the critiques we have mentioned so far by dismissing the assumption that for a given number to become an adequate indicator of well-being, this number has to be philosophically grounded and empirically valid in the light of the best available standards in philosophy, measurement theory, and psychology.

In the academy, both the cheerleaders of life satisfaction and its critics go along with this assumption. For the cheerleaders, it is an ideal that provides intellectual legitimacy to the whole enterprise, though they add that the bar for validity should not be set too high. For them, life satisfaction meets a certain minimal theoretical standard—it is after all not completely ridiculous to suppose that well-being is a matter of how well we judge our life to be going. The critics do not buy this and demand a rigorous justification of the very idea of well-being quantification. Philosopher Daniel Hausman (2015), for example, rejects all the existing measures simply because well-being is too complex and too person-specific to be captured by any population-level scale. For both sides, however, the matter is intellectual.

For the political sponsor, however, the arguments that measures of well-being do or do not get at "the real thing" miss the mark. Science and policy are not after the real thing, but seek a redefinition of well-being that, while preserving something of the parent-concept, picks out a phenomenon in the world that is measurable, rendering it comparable and transferable. This is the argument that comes out most clearly in a testimony by Oliver Letwin, the minister for government policy under David Cameron and the latter's influential adviser. As the Conservative government came to power in 2010, well-being was a centerpiece of its agenda for the "social revival" of the United Kingdom. Ironically, this agenda was coupled with austerity measures, and was arguably cast in the role of a warm and fuzzy distraction from the slashing of public budgets.[9]

In 2016 and now out of power, Letwin recollected the challenge of a politician interested in well-being thus: "If you talk in a debate about things like beauty, or happiness, or life satisfaction, or well-being, it does not take more that fifteen seconds before you are dismissed as an eccentric lunatic" (Centre for Economic Performance 2016). This, he continues, is in sharp contrast with values that are nicely quantifiable such as "weapons, hospitals, rail lines." Letwin and his Tory colleagues wanted to change the terms of this conversation by making well-being a legitimate subject to be brought up at town-hall meetings, media interviews, and election debates. Letwin freely admitted that this required representing well-being as a quantity even though in fact, according to him, it is a quality. (Letwin has a PhD in philosophy, hence his comfort with these distinctions.) And it required not just saying that well-being can be represented quantitatively but also institutionalizing this view with statistics collected by dispassionate bureaucrats, reports, graphs, and other essential accouterments of authority. Thus was born the Measuring National Well-Being program in the Office for National Statistics (ONS), which since 2011 has continuously collected relevant statistics and regularly issued reports on well-being along with other uncontroversial numbers like crimes, births, and so on. Letwin's objective was to obtain "a set of data which may be naïve, but also respectable and internationally comparable, and [which] can be used in political debate."[10] For bureaucrats and politicians like him, measurable and quantifiable well-being became an essential ingredient of policy debates.

The conflict Letwin describes—"Letwin's dilemma"—is real and generalizes beyond this episode of British political history. Well-being numbers are artificial and distort the real thing, and they can have unforeseen consequences; but without these numbers public debate focuses only on conventional economic indicators, which have these failings too. As a political sponsor of well-being measurement, Letwin clearly picked one horn of this dilemma. He set in motion the operation that now reliably produces numbers about British national well-being and that we describe in the next section.

Let us take stock of the argument so far. Quantifying well-being is a messy business. There is no unique and obviously correct definition and any choice of indicators will be controversial—even the most established measures such as life satisfaction. But there is a strong practical impetus to produce some credible numbers about well-being. Many scientists are enthusiastic about challenging what they perceive as economics' unfair domination in policy; politicians like Letwin are motivated by the desire to show a commitment to fundamental values and to redirect public spending accordingly. With this story in hand, we are in a position to move into the more critical territory. Well-being measurement always requires a compromise. Are some compromises better than others?

Let us first preempt one possible reaction—that of rejecting the framing of the dilemma entirely: "Who says that the subjects of political debate and public policy must be quantitative? Why can't well-being be part of the conversation without being numerical? Who gets to set these terms and why?" We recognize that there is room for different argumentative strategies and, as Chatterjee and Newfield demonstrate in their respective chapters in the present volume (chapters 1 and 2), the numerical can, as a matter of fact, be dislodged. Still, we agree with John (chapter 6) and Badano (chapter 7) that it is too rash to dismiss the advantages of quantification in democratic politics. When politicians make numerically precise promises, it is easier for the public to assess how they deliver on those promises. Numbers provide a means of speaking truth to power for those who are less articulate with compelling narratives and beautiful stories. Statistics may be fickle, but numbers make debates concrete, and can be disputed on scientific grounds in a way that qualitative concepts and narratives cannot. So the dilemma stands.

Which horn is preferable? Is it better to settle for distortion or for irrelevance? We submit that there is no straightforward resolution of this dilemma. There is no argument powerful enough to show that well-being numbers are always preferable to other numbers, nor that they are always inferior. Rather it depends on the specifics and on the context: *which* well-being measures are used, what they are used *for*, and what alternatives are available. To illustrate the complexity of the matter, we will consider two examples from public policy. In our view, the first one represents a defensible generation of numbers about well-being, while the second less so.

8.5. INCORPORATING WELL-BEING INTO NATIONAL STATISTICS

The ONS project to monitor the United Kingdom's national well-being spurred a regular production of rich and diverse statistics describing all relevant spheres of life. (Indeed unlike almost everything else Cameron's government did, the well-being measurement initiative received no criticism from other political parties.) Its main virtues are its comprehensiveness and its legitimacy. Both are a result of the seriousness with which the ONS approached the task of drafting an inclusive list of indicators that are meaningful to the public. To secure this, the ONS conducted a countrywide consultation called "What Matters to You?" between 2010 and 2012, soliciting views and recommendations from the public, experts, and communities all across the United Kingdom (Office for National Statistics 2012). Potential measures of well-being were released to the public and then respondents were queried about their suitability:

- Do you think the proposed domains present a complete picture of well-being? If not, what would you do differently?
- Do you think the scope of each of the proposed domains is correct? If not, please give details.
- Is the balance between objective and subjective measures about right? Please give details.

The outcome of this exercise is a measure that contains both subjective indicators—happiness, life satisfaction, sense of meaning—and objective indicators, such as economy, work, health, education, safety, housing, and recycling (Office for National Statistics 2019). The ONS settled the seemingly intractable debates involving the experts and various groups of the public by including as many items in its final measure as practically possible and also by publicly vetting this measure. Their approach is similar to schemes in France, Italy, Canada, and New Zealand, as well as the German initiative "Gut Leben in Deutschland," which used town-hall discussions in 2013 to arrive at a list of twelve groups of indicators against which government policies were to be judged. (Interestingly, none of them concerned subjective well-being.)

This model of producing well-being statistics can be criticized using both horns of Letwin's dilemma. The table of indicators is rich in information, thus reflecting the complex and pluralistic nature of well-being, but this very richness hides deep disagreements about which indicators are more central than others. Is access to recycling on the same plane as freedom from anxiety? How do all these indicators work together? Is ONS National Well-Being "everything but the kitchen sink"? Whether these questions are answerable to the critic's satisfaction or instead expose fundamental problems depends on how the data collected will be used. When Cameron's cabinet kick-started the project, they had great ambitions for judging policies and spending priorities against the ONS data. This did not come to pass, since after the Brexit referendum the well-being agenda died, at least in the upper echelons of the government. Now the good work of the ONS serves mostly a representational function: their state-of-the-art statistics paint a multifaceted picture of a community's life according to standards that this community itself endorses. Granted, the picture could be even richer. For example, the ONS could collect qualitative data about well-being in the form of narratives and interviews. They could also take a cue from the growing movement of citizen science and more systematically involve citizens in generating data that they see as reflecting their well-being. But this might be too much to ask from a government department. So given the ONS position and constraints, their statistics do the job. In this sense the ONS found a defensible middle road out of Letwin's dilemma.

8.6. LIFE SATISFACTION AS THE MASTER NUMBER

A very different project is undertaken by economists who wish to do a lot more than just reflect the variety of well-being-related priorities in national statistics. To them it is not enough to just measure well-being, they want to incorporate it into the very heart of policy evaluation—the cost-benefit analysis—so that well-being becomes the benchmark against which each item of public spending is judged. For this to happen, tables of varied indicators like those collected by the ONS are entirely unsuitable. Rather, these economists are after a single quantity whose variation in response to changes in policies and circumstances can be observed, thus enabling the identification of bundles of policies that maximize overall happiness.

The economists advancing this vision most explicitly are Andrew Clark, Sarah Flèche, Richard Layard, Nattavudh Powdthavee, and George Ward, of the Wellbeing Programme at the LSE's Centre for Economic Performance, whose manifesto was published as the 2018 book *The Origins of Happiness: The Science of Well-Being over the Life Course*. There are others who subscribe to this vision (De Neve et al. 2020; Frijters et al. 2020), but our main focus will be on the *Origins*. The book received ringing endorsements from the most prominent well-being scientists in the United States, Canada, and elsewhere in Europe, and in December 2016 a high-profile launch event was held at the LSE to unveil the key findings of the project. It was attended by national and international powerbrokers and widely covered by the media.

It was at this event that Letwin articulated his dilemma, and it was clear which horn the authors of *Origins* preferred. By 2016, Richard Layard had spent decades popularizing the science of well-being and devising ways to deploy it in policy. He is explicit in endorsing the utilitarian goal of furthering subjective well-being—and for that, life-satisfaction data are plenty good enough (Layard 2005). None of the objections against them, or the agenda as a whole, appear significant to him. When introducing *Origins* at the launch he briefly went over the worries about validity of life-satisfaction reports, retorting that they correlate well enough with brain scans and longer questionnaires. Even more briefly he noted that some critics find that there is more to life and good governance than happiness—namely justice, rights, and fairness. However he dismissed these worries as "puritanical." The strategy of the book's authors appears to emphasize the strength and high profile of their allies rather than to engage with their academic detractors. For example, they dedicated the book to the patron of happiness economics in the UK policy world, Gus O'Donnell; gave a shout-out to Tony Blair in a chapter epigraph; and, finally, invited Ohood Al Roumi, the minister of state for happiness from the United Arab Emirates, to share her experiences implementing happiness policies there. When a troublemaker in the audience asked about rights of foreign workers in the UAE, she inevitably had to ignore him. The overall impression of the authors of this chapter, who were present at that launch event, is that for Layard and his team the moral and intellectual complications were a small price to pay for the potential practical significance of their conclusions.

The heart of their proposal is that the data available from various national and international panels (such as the British Household Panel Survey, the German Socio-Economic Panel, the Household Income and Labour Dynamics in Australia, and the Avon Longitudinal Study of Parents and Children) enables fairly precise inferences about how much a given set of social, demographic, and economic circumstances boosts or impedes happiness, defined as life satisfaction. And these inferences can be made over the life course, starting from childhood all the way to old age. For adults, mental health (self-assessed) emerges as the single biggest predictor of individual happiness, and similarly for children—though in their case it's assessed by the mother, and the mother's own mental health is the best predictor of the child's. Other factors are also considered, such as poverty, education, parenting styles, school, employment, partnership, social norms, and so on, but none makes as strong a statistical contribution as mental health. For example, having a diagnosed depression or anxiety disorder explains twice as much variation in life satisfaction as income ($R^2 = 0.19$ for mental health and only 0.09 for income). A person's education has even less effect than income ($R^2 = 0.02$), whereas the education of others in one's surrounding has a measurably *negative* effect on individual life satisfaction. This is part of a well-documented effect known as "social comparison," where the self-estimated value of your income or education depends crucially on how much of those goods others around you possess. Another phenomenon such analysis reveals is adaptation, that is, returning to the previous level of life satisfaction after a positive or negative shock. This, however, does not hold for unemployment, loss of partner, and mental illness.

The authors document these facts with great care, reporting the relative and the absolute quantitative effects of each specific factor in life satisfaction using coefficients. These coefficients play an essential role in the "revolution in policymaking" that Layard and his colleagues advocate. They argue that government spending should be evaluated pretty much exclusively using a method of cost-effectiveness in which benefits are measured in units of happiness. In this vision, each item of government spending must pass a test: does it increase happiness as efficiently as possible? Such an exercise requires a threshold of cost-per-unit-of-happiness, below which programs and services should not be funded. The authors see quality-adjusted life years (QALY), currently used by the National Institute for Health and Care Excellence (NICE) and discussed also by Badano (chapter 7), as the obvious model. As Badano explains, NICE recommends against public provision of drugs that cost more than £30,000 per QALY and this number, although "spurious" in John's sense (chapter 6), plays a legitimate political role. Layard and coauthors propose to extend this process from health to well-being. They hypothesize that it is not efficient for the Treasury to recommend spending that costs more than, say, £3,500 per unit of happiness. This is why the estimation of absolute effect coefficients is so important to the authors of *Origins*. Once we know how much extra happiness income, unemployment, health, or what have you, buys, we will be able to compare the cost-effectiveness of different policies. Public moneys will go toward the happiest possible bundle of services.

What are we to make of this example of well-being quantification? Commentators from different fields will raise different objections. The critics of life satisfaction discussed in section 8.4 will worry about the exclusive reliance on this indicator (none of the other well-being indicators collected by the ONS features in *Origins*). Ethicists and political philosophers, on the other hand, will ask how rights, obligations, and constitutional constraints will feature in the proposed cost-effectiveness analysis (see Fabian 2018). But we want to focus on the specific problem of the use and misuse of numbers. In *Origins*, numbers serve to erase social, cultural, and historical context and to turn well-being into a simple object with universal determinants discoverable by statistics alone. Badano and John each make a strong case that spurious numbers may play legitimate political roles. But we doubt that such a justification is available in this case. Two instances of context erasure are particularly vivid: how the authors treat mental health and how they treat public goods.

In *Origins*, mental health is measured largely by brief standardized self-reports, and these reports, it turns out, explain a large chunk of variation in life satisfaction, more than poverty and inequality do. So the authors tout as their big result the idea that mental illness is the biggest cause of misery, and that intervening on it, rather than on poverty, is the most efficient way of raising happiness. An obvious circularity arises here because questions about life satisfaction are very similar to questions of self-reported mental health. But more significantly, as the network of activists Psychologists for Social Change argues, had the authors used more than simple regression modeling—had they attended to the rich tradition of qualitative research in this area—they would not treat mental health and poverty as noninteracting variables that each make a separable contribution to well-being, one large and one small. Just as intersectional feminists worry that gender and race cannot easily be decomposed into distinct causes of oppression, so poverty and mental illness should be studied together as coproducers of misery by a combination of qualitative and quantitative methods. In this case, *Origins* puts forward a number with questionable validity and misleading precision and then uses this number to support the consequential conclusion that poverty should be less of a priority to policymakers than mental health.

Likewise, in the analysis of the relationship between public goods and happiness, *Origins* seeks to explain variation in happiness across 126 countries using social variables such as trust, generosity, freedom, and social support. A cross-sectional regression based on Gallup World Poll data seemingly provides enough evidence for the authors to make universal causal pronouncements such as, "If we go from the lowest levels of trust (7% in Brazil) to the highest levels of trust (64% in Norway), this raises average life satisfaction by 57%" (Clark et al. 2018, 229). Our worry here is not the limitations of this exclusively cross-sectional analysis (that it fails to tackle the question of whether trust influences happiness or happiness influences trust). Nor is it that the trust variable is defined narrowly as the proportion of people who say "yes" to the single question, "In general, do you think that most people can be trusted?" Rather, the problem is the assumption that well-being determinants have a universal noncontextual effect that coefficients measure: x for mental health, y for education, z for trust. Of course, it is possible to disaggregate the statistics by populations and hence recognize the differences (which many happiness economists do). But recall that the authors of *Origins* are seeking a number relative to which policies should or should not be funded. To settle on such a number you need to treat the effect of a variable identified by statistical analysis as a uniform contribution of this variable always and everywhere. This uniformity is extremely implausible. The strong statistical effect of mothers' mental health on children may have something to do with the gender politics of the surveys, or it may stem from the fact that mothers did most of the care work at the time when the data was available. Might fathers' mental health become equally important as they do more of the childcare? Why present this effect as a stable cause of child well-being, as if mandated by nature? The data reveal plenty of fascinating differences among countries in the way that, say, loss of employment or disability affects life satisfaction, but these contextual effects do not sit well with the ambition to estimate quantities that can be plugged into cost-effectiveness analysis. The authors' lack of interest in the local, the variable, the historical is motivated by the self-imposed demand for uncontroversial, stable, and tractable input into cost-effectiveness analysis. In this project, as they see it, there is no room for qualitative data, for rich but local ethnographies, nor indeed for participatory policymaking.

This example of generating and using well-being numbers, we think, is very different from the multitudinous and evolving tables of indicators produced by the ONS, even considering that those also lack qualitative information. The project of the *Origins* is far more controversial, because far more audacious in the way it transforms well-being into an object of quantification. We see here a metamorphosis of life satisfaction from an academic indicator that challenged economists into a master number that has stable determinants measured by coefficients. This is in stark contrast with the conclusions of the aforementioned Stiglitz, Sen, and Fitoussi report so often cited as an inspiration for happiness economics. That report's vision of well-being is nothing like that of *Origins*: it argues, for example, that "no single measure can summarize something as complex as well-being" (Stiglitz et al. 2009, 12). Perhaps this recognition is nothing more than lip service, similar to Letwin's blithe remark that well-being is "of course" a quality rather than a quantity. Perhaps once quantification of well-being gets underway, there is an inevitable tendency toward its reduction to the master number of life satisfaction. But these possibilities should not stop us from ringing alarm bells.

Letwin's dilemma invites us to imagine just how bad it would be to use "naïve" well-being numbers as compared to standard economic indicators. We can't make a watertight case as to which is worse: the proposal in the *Origins* or the existing model of representing benefit and evaluating policies? We have given some reasons to think that Layard and coauthors distort well-being in the direction that makes it hardly recognizable and do so without properly engaging with their critics. In their hands, well-being becomes a monistic quantity that reacts mechanically to changes in circumstances as if it were a Newtonian system with forces that combine by vector addition. Even those open to quantification of well-being for the sake of democratic policy deliberations, along the lines of John and Badano, should balk at such a radical transformation. Well-being may well be a pliable concept, but is it that pliable?

8.7. ADVICE FOR THE CRITIC

What lessons do our two stories carry for the task of analyzing quantification more generally? We emphasized that when it comes to quantifying well-being there is no master measure. There is, however, a very popular measure—life satisfaction ratings—and this quantity, in virtue of being easily available and unidimensional, has made it further than other measures into the world of evidence-based policy. Whether this is a good thing is not so much a question of whether it, or any other measure, represents well-being properly, but rather of what numbers we compare it to. Well-being numbers in the newly updated national statistics (which include life satisfaction among other indicators) seem a huge improvement over the limited and narrow previous data. However, in the hands of the authors of *Origins* urging a new form of cost-effectiveness analysis, the notion of "life satisfaction" is much less innocent because it is paired up with a grossly implausible methodology of social evaluation as well as a larger technocratic model of governance.[11]

Although we concentrated only on public policy, such pairings between more and less controversial can also be found in the use of happiness numbers in self-help and management. So our guess is that the lessons generalize. Responsible criticism has to take seriously consideration of validity—that is, whether the measure in question is an adequate representation of well-being. For this purpose we need to wear the hat of a philosopher-scientist who articulates and endorses a certain minimal standard of well-being measurement. Yet this is not enough. The philosopher-scientist then needs to put on the hat of a social scientist who acknowledges the rhetorical and pragmatic role this number plays in politics, governance, and public debate, and judges how well it plays this role as compared to other numbers. The critic should be ready to accept certain trade-offs between usefulness and validity, because there is no point in holding well-being measures to an impossible ideal. The scope of considerations justifying these trade-offs should be wide, encompassing the moral and political work these numbers do.

NOTES

1. Diener and Seligman (2004); Huppert et al. (2005); Kahneman and Krueger (2006); Kahneman et al. (1999); Layard (2005); Seligman (2004); Seligman and Csikszentmihalyi (2000).

2. This table is an abbreviated version. For a full version and references, see Alexandrova (2017).

3. For example, development economist Angus Deaton describes his Nobel Prize–winning research as concerning "wellbeing, what was once called welfare, and uses market and survey data to measure the behavior of individuals and groups and to make inferences about wellbeing" (Deaton 2016, 1221).

4. Stevenson and Wolfers (2008) articulated an influential critique. Clark at al. (2012) present the state of the art, and OECD (2013) defends the continued relevance of well-being research.

5. Kahneman et al. (2004a, 2004b) explain the virtues of experience sampling for science and national accounting; the work of Stone et al. (2016) is an example of its use in research.

6. There are different ways of measuring life satisfaction. The Satisfaction with Life Scale (SWLS) is one popular five-item Likert scale. Its prominence is due to it being short and to the fact that it has by now been used in hundreds of studies, especially by its originator, the prolific psychologist Ed Diener and his colleagues and students (Diener et al. 1985, 2008). Questions about life satisfaction also figure in all the main large-scale surveys and panel datasets worldwide, such as the German Socio-Economic Panel, the UK British Household Panel Survey, and the Australian HILDA Survey.

7. Though see Larroulet-Philippi (2021) and Vessonen (2019) for worries about these standard conventions regarding validity.

8. See Lucas and Lawless (2013) and Oishi et al. (2003) for a defense of life satisfaction judgments, and Deaton and Stone (2016) and Lucas et al. (2016) for the latest debate on context effects.

9. See Davies (2015) for a critical commentary. For a sympathetic one, see Express KCS (2015).

10. Centre for Economic Performance 2016.

11. We develop this argument further in Singh and Alexandrova (2020). For more on whether the practicality of the life satisfaction measure justifies its dominance, see Mitchell and Alexandrova (2020).

CHAPTER 9

Aligning Social Goals and Scientific Numbers

An Ethical-Epistemic Analysis of Extreme Weather Attribution

Greg Lusk

People increasing rely on social technologies to make all sorts of social de-cisions. Examples from daily life abound: smartphone apps estimate the density of cars and their speed in order to detect traffic and suggest time-saving routes; websites compare home prices and neighborhood tax infor-mation to estimate values for would-be buyers; and fitness trackers count our steps, swimming strokes, and hill climbs against our personal health targets, automatically posting successes to social media. The algorithms, and the quantitative data they rely on, give users knowledge that previ-ously required consulting the traffic desk of the local radio station, real estate agents, and personal trainers. Armed with this knowledge, users can make decisions "for themselves," which is often equivalent to trusting such technologies to decide for them.

Of course, the desire for easily understood information that could en-hance decision-making is not reserved for individuals driving to the air-port or buying houses. In his influential history of quantification, *Trust in Numbers* (1995), Theodore Porter locates the prestige of quantifica-tion in its ability to support social technologies that tame subjectivity and enhance managerial power. As he sees it, the widespread emphasis on numbers in public spaces is not likely to be merely a case of "physics envy": businesspeople, administrators, and government officials use social technologies like cost-benefit analysis to assert control over the decision-making process by reducing reliance on experts. Quantification helps enable such technologies, in turn enabling the managerial and adminis-trative class to make decisions for themselves, displacing once-necessary expertise. Viewed in this way, quantification, and the processes that uti-lize it, plays a central role in shaping social decision-making by providing information that foregoes subtlety and depth to achieve usefulness.

From Porter's work, it is easy to see why the Original Critique of numbers in decision contexts—in short (and as outlined in the introduction to this volume) that the seductiveness of numbers obscures nonquantifiable aspects of phenomena—was so powerful. Not only did certain methods of quantification render invisible that which could not be expressed by numbers, but those at the helm of this process were often those with substantial power. As a consequence, other methods of quantification were sometimes quelled, forgotten, or ostracized, along with the less powerful who might have benefited from these different numerical approaches. Given this, it seems unsurprising that there is now a growing populace skeptical of experts and their currency of quantification in the United States and United Kingdom. Some within this skeptical movement feel left out of the equation. This skepticism makes clear what has always been true, but was not always emphasized in the Original Critique: Who or what is quantified is always a matter of decision. Quantification itself is neither virtuous nor vicious; the character of numbers depends on how they are wielded. Good numbers can be useful and bad ones can easily mislead or obscure.

How do we make and use numbers in a responsible way? This chapter begins to address that deceptively simple question. I will show that there is a central insight shared between those who articulated the Original Critique (using Porter as an exemplar) and certain views of measurement in philosophy of science. The shared insight is that quantification is itself perspectival. Numbers—despite claims to objectivity—always carry with them a certain orientation toward that which they represent. This orientation connects the ethical impacts of quantification to the epistemic perspective that such numbers promote. In such a case, then, one sign of a virtuous method of quantification would be alignment between quantification's representational capacities and laudable social ends. Achieving this alignment is harder than it may seem, as I show through a coupled ethical-epistemic analysis of extreme weather attribution. Extreme weather attribution is a developing methodology in climate science that proponents argue is a promising social technology that can be used to spur adaption and climate justice. This analysis shows how and why certain numbers may or may not align with their intended purposes, but also how one might begin to assess the virtues and vices of quantification that bridge science and decision-making.

9.1. QUANTIFICATION AS A PERSPECTIVAL PROCESS

Porter (1995) advances two theses in an attempt to explain the prestige and power of quantification in the modern world. The first thesis claims that the arrow of explanation runs counter to what might be intuitively supposed: the appeal of quantification to businesspeople, administrators, and government officials explains aspects of numerical reasoning in the natural and social sciences, rather than the other way around. The second thesis claims that quantification smoothes over the need to perform deep and nuanced analysis in a way that allows relatively unskilled persons to make decisions that functionally replace expert judgment. Following Ed Levy (2001), I will call the first the *reverse transfer thesis* and the second the *judgment replacement thesis*.

For Porter, quantification often serves as a social technology that drives a practical imperative he calls "the accounting ideal" (1995, 50), which is crucial for the management of people and nature. Porter explores this ideal by examining historical episodes, most notably through a comparative examination of the adoption of cost-benefit and risk analysis by French state engineers and the US Army Corps. However, one simple example, which Porter (47–48) borrows from William Cronon (1991), demonstrates how the accounting ideal furthers judgment replacement: prior to 1850, wheat in the Midwest United States was shipped via river in bushel-sized sacks that differed in weight due to the density of how they were packed. Each of these sacks was examined individually by a miller or wholesaler in order to establish its value. There was at this time hardly any common notion of the "price of a bushel of wheat," or if there was, it was known only to those who routinely established its value. But by 1860, the Chicago Board of Trade had defined the bushel in terms of a standard weight and divided wheat quality into four separate grades. Once quantified, wheat could be bought and sold on the Chicago Exchange by those with comparatively little knowledge about wheat production or quality. Porter sums up the episode nicely: "The knowledge needed to trade wheat had been separated from the wheat and the chaff. It now consisted of price data and production data, which were to be found in printed documents produced minute by minute" (1995, 48). No longer was the judgment of a miller needed to price wheat; the locus of wheat-valuing expertise had shifted to traders and investors.

Quantification replaces judgment, on Porter's account, by establishing a set of rules that enable "mechanical objectivity." Such rules displace the subjective judgment and beliefs of individuals from the process of social management, replacing them with recipes for quantifying objects or processes of interest. When a community is small and its members trust each other, quantification is largely unnecessary. What quantification provides, via mechanical objectivity, is the capacity for management at a distance: numbers become tools of communication whose objectivity establishes their authority and usefulness. Numbers produced according to rules allow those without expertise to engage in practices that previously required expert judgment, as the wheat example above shows. Porter notes that when quantification successfully proceeds in this way, it "nearly always comes at some cost to subtlety and depth" (1995, 86).

Thinking about quantification as a means of refining the social division of labor through the displacement of judgment, as Porter does, is not traditionally the approach taken in philosophy of science. The default approach there, if there is one, is to view numbers, when produced by trustworthy measurements, experiments, and computer simulations, as representing phenomena in a way that, at least under certain conditions, allows them to be shared as empirical evidence among researchers.[1] Thus philosophers have been much more interested in characterizing the general inferential patterns—the judgment that is often replaced on Porter's account—that result in successful mathematical representation of physical or social phenomena. The power of quantification, for many philosophers, is located within the deep and nuanced reasoning that produced particular numbers that successfully reflect phenomena out in the world. As Ed Levy (2001) points out, quantification is often the result of rigorous expert judgment, not a lack of depth and subtlety.

However, these two views of quantification are more complementary than critics like Levy would make them seem. After all, there is a great deal upon which many philosophers of science, and Original Critique historians like Porter, agree. Take, for example, the newly revived philosophy of measurement, whose goal is to provide an epistemology for the practice of representing phenomena by numbers (an area long neglected by philosophers of science). One of the lessons emerging from this literature is that measurement "provides a representation of the measured item, but also represents it as thus or so" (van Fraassen 2009, 180). For example, when measuring the temperature of a cup of tea, the measurement represents the tea as it "looks" from within a particular measurement setup in a particular measurement environment. What vantage point to look from—at least before the kind of standardizing norms that result in mechanical objectivity are established—are largely left to the judgment of scientists. Measurement, and in fact numerical representation writ large, is thus a selective form of representation: how the represented object is quantified will be a result of the way which it is "viewed."

Where the Original Critique of quantification reflected in Porter's work and the representational approach typical of philosophy of science come together is precisely on this point regarding the inevitability of perspective: numerical reasoning always has a specific vantage point. Of course, the vantage point emphasized in these literatures is different. Proponents of the Original Critique emphasize the vantage point of social actors who use quantification to shift the locus of judgment, while philosophers of science have emphasized how numbers representationally carry with them a particular vantage point based on the decisions of their creators. Regardless, both admit that different vantage points may lead to different judgments. And while proponents of the Original Critique and philosophers of science acknowledge the important role of mechanical objectivity, mere adherence to rules of quantification does not remove the influence of perspective. The rules that underwrite mechanical objectivity encode the choices that give numbers their perspectival character. In many ways, these complementary perspectives are two sides of the same coin.

When answering normative questions about when it is permissible to allow numbers to guide social decisions—at least those numbers that purportedly represent physical phenomena—a dual analysis therefore seems to be in order. One needs to examine both the goals that such decisions aim to achieve and whether the kinds of quantifications involved are fit to achieve those goals. Thus, we can say that there needs to be alignment between two sets of vantage points—those that guide number creation, and those that guide number use. Setting aside for a moment the questions regarding the desirability of goals, this kind of analysis serves as a minimal criterion that can help establish whether certain kinds of numbers are adequate for the purposes for which they are often used.[2]

In the sections that follow, I engage in a coupled ethical-epistemic analysis of extreme weather event attribution by way of demonstrating one approach we might take to normative questions about number use. Such analyses are a relatively new way of linking scientific research with social outcomes to help determine how research should be used (see Tuana 2013 and Tuana et al. 2012 in the context of climate science, and Katikireddi and Valles 2015 for a biomedical application). Coupled ethical-epistemic analyses aim to connect the methodological and epistemological aspects of scientific research with the ethical consequences that research has when it is used socially, particularly in decision-making. Such analyses acknowledge that science—despite its aim of objectivity—often involves the use of contextual and social values within its practice, especially when science aims for public relevance. Put in the vocabulary of this section, the choice of quantitative perspective is often a value-laden one that has consequences for the way research should be used when addressing social problems. Coupled ethical-epistemic analyses are useful because of their dual character. I will deploy this kind of analysis here to demonstrate both that the epistemic choices made in quantification often have ethical consequences, and that our ethical or social choices should inform our means of quantification.

9.2. THE CASE OF EXTREME WEATHER ATTRIBUTION

There is little doubt among climate scientists that the observed trend toward a higher global mean temperature is largely the result of human actions (Cook et al. 2013; Oreskes 2004). Anthropogenic forcings, dominated by the release of greenhouse gases as well as land-use changes, are responsible for altering the Earth's radiation budget by trapping a larger amount of thermal radiation in the atmosphere. This trapping of thermal radiation—that is, heat—is commonly known as the greenhouse effect. Less heat leaves the planet's atmosphere, resulting in higher global mean temperatures. As the latest Intergovernmental Panel on Climate Change (IPCC) report indicates, there is an unequivocal human influence on the climate system that has likely already committed the planet to at least a 1.5°C rise in mean temperature (Stocker et al. 2013).

While scientific investigations of climate change often take place on the global scale, scientists are equally interested in the repercussions of an altered atmosphere on regional and local scales. One area of significant concern is extreme weather, such as heat waves, cold snaps, droughts, deluges, and hurricanes. These can cause large-scale destruction of human infrastructure, resulting in costly physical damage and the breakdown of food systems and supply infrastructure. Extreme weather event attribution attempts to do for these events what scientists have already done for the rising trend in global mean temperature: show that they are influenced by human actions.

Such attribution involves assessing how the properties of extreme events have changed given anthropogenic influences. Often the change that scientists are interested in is the frequency of occurrence of a particular type of event in a particular place. The type of event is defined by threshold exceedance in a specific region of the globe, for example, the record high July mean temperature in western Europe. Scientists may use the old record high temperature as a threshold to assess how the probability of that record being broken has changed due to anthropogenic forcings. This form of event attribution, known as *probabilistic event attribution*, involves establishing the frequency of occurrence of a particular type of event in the "natural world," where anthropogenic forcings are absent, and comparing it to the frequency of extreme event occurrence in the "actual world," where anthropogenic forcings are active. The difference in the frequency of occurrence between the two worlds can therefore be blamed on the presence of anthropogenic forcings. This "blame" is the attribution: one attributes the observed increase in probability of occurrence to anthropogenic factors. This is a risk-based approach, in that it says how much more likely a type of event is given anthropogenic climate change (i.e., how much more "risk" of occurrence there is).

There are at least two methods of computing the frequency of event occurrence in the natural and actual worlds. The first might be referred to as the "classical statistical" approach (Hulme 2014) and the second the "model-based" approach. In the statistical approach, time series are used to detect outliers or changes in trends. For example, Luterbacher et al. (2004) reconstruct European monthly and seasonal temperatures from a host of sources, including proxy data. The reconstruction is used to demonstrate that summers like that of 2003 are outliers, and 2003 was likely the warmest European summer since 1500. Scientists often use these reconstructions to calculate the return period of such events. Luterbacher et al. calculate the return period for a 2°C summer anomaly at millions of years for early twentieth-century conditions and less than 100 years for recent summers (2004, 1502). The implication is that the change in return time over the twentieth century—that heat waves of this kind are returning significantly more frequently—should be credited to anthropogenic global warming.

Another means of computing these frequencies involves the use of computer models to calculate the changed odds of an event. This approach involves running one set of computer simulations of the climate system (or a subregion of that system), which explicitly represents anthropogenic forcings, and another set with the same structure except with the anthropogenic forcings removed. Scientists use the products of these two sets of simulations to estimate the probability of a particular kind of event in the current climate, and in a nonanthropogenically altered climate. They then compare these two probabilities and compute what is called the "fraction of attributable risk," or FAR. This quantity tracks how much of the current risk of event occurrence is due to anthropogenic forcings; for example, if FAR = 0.5, then half of the current risk of the event occurring is due to anthropogenic factors or, put another way, the probability that an event of a defined type will occur has doubled.

It is important to note that probabilistic attribution, regardless of the approach, does not attribute an actual weather event to anthropogenic factors. In fact, it is admitted that any particular extreme could have happened completely naturally. Probabilistic event attribution attributes only *the increased probability of occurrence* to anthropogenic factors. Thus, one cannot "blame" *any particular occurrence* of an individual extreme event on anthropogenic forcings; one can attribute only an *increased chance* of a particular *kind of event.*

It is also important to note that probabilistic event attribution is not forward-looking, and says nothing about the future. Probabilistic event attribution merely compares how the current risk of an extreme event differs from the risk of an event in the natural scenario (or, in the case of the classical statistical method, some prior point in time when anthropogenic forcings were negligible). It says nothing about how the future risk of extreme events might differ from today's risk. After all, one cannot attribute something that has not yet occurred. Nonetheless, scientists want to use the practice to garner trust among the public and decision makers by replacing the sometimes contentious expert judgments of climate scientists.

9.3. EXTREME WEATHER ATTRIBUTION AS A SOCIAL TECHNOLOGY

The case of extreme weather attribution is interesting because it has been deliberately positioned as a social technology that will support social decision-making. As a National Academy of Sciences report indicates, "the primary motivation for event attribution goes beyond science"; it is to provide valuable information to emergency managers, regional planners, and policymakers who need to grapple with a changing climate (National Academies of Sciences, Engineering, and Medicine 2016, ix).

Most of this positioning has been done by a small group of probabilistic event attribution proponents affiliated with the University of Oxford and the United Kingdom's Met Office, most notably Myles Allen (Oxford), Friederike Otto (Oxford), and Peter Stott (Met Office). They seemingly subscribe to Porter's reverse transfer thesis: they have argued that extreme weather attribution is valuable because it would allow for enhanced management of the impact of global warming by those outside of climate science. As Myles Allen has stated, "Because money is on the table, it's suddenly going to be in everybody's interest to be a victim of climate change. . . . We need urgently to develop the science base to be able to distinguish genuine impacts of climate change from unfortunate consequences of bad weather" (quoted in Gillis 2011). In other words, because decision makers are going to face a whole host of questions about how to distribute resources in a changing climate—including questions about compensation—we should adopt scientific strategies of quantification so that those decisions can be made efficiently and fairly. This message of social usefulness dominates discussions about extreme event attribution; scientific benefits are scarcely mentioned.

Allen and Stott recognize that in order for event attribution results to replace specialized judgments for the purpose of social management, the results need to be seen as objective. Allen and Stott argue that probabilistic event attribution could constitute a "relatively objective" approach to extreme weather so long as an "industry standard" methodology—like the one they propose—is adopted (Allen et al. 2007). In a strategy straight out of Porter's playbook, they argue that adopting such a standard would eliminate bias and minimize the need for expert judgment.

Allen and his colleagues want to avoid inaction due to dueling experts: climate change is a contentious issue, and at least on the finer details, the experts sometimes disagree. Such disagreement is often seen by the public as a sign of more general uncertainty, which leads to social inaction. By adopting an industry standard—or as Porter puts it, instituting some form of mechanical objectivity—Allen and colleagues seek to replace expert judgment by algorithmic calculation. Unsurprisingly, the judgment they want to retain is their own, as encoded in the very form of probabilistic event attribution they developed. In essence, probabilistic event attribution packs the judgment of a subset of scientists into a numerical figure to enable social decision-making.

Scientists, particularly those working in the spirit of Allen and Stott, have touted numerous areas of social need that would benefit from extreme weather attribution. The following sections examine the fitness of probabilistic event attribution for the social ends which motivate its development. Specifically, I examine if extreme event attribution can meet the following goals: (1) provide information about the future risks of extreme events to emergency managers, regional planners, and policy makers; (2) help society build adaptive capacity and ensure proposed adaptations target the right kind of events; and (3) attribute particular events so that victims of anthropogenic climate change can be recognized and distinguished from victims of naturally extreme weather—cases of bad luck—and perhaps also be compensated.

9.4. PROBABILISTIC EVENT ATTRIBUTION: AN ETHICAL-EPISTEMIC ANALYSIS

The need to ensure alignment between the means of number production and the proposed or actual use of such numbers is evident when considering goal 1 above, the use of probabilistic event attribution to inform decision makers about the future risks of extreme weather. There seems to be a conceptual confusion within the broad scientific community and science journalists regarding the character of event attribution studies that could mislead potential consumers of the information. For example, the National Academies of Sciences, Engineering, and Medicine (2016) make a claim that is often repeated in popular conversations about event attribution: a "defensible analysis of the attribution of extreme weather events . . . can provide valuable information about the *future risks* of such events to emergency managers, regional planners, and policy makers" (ix; emphasis added). Similarly, Stott et al. (2013) suggest event attribution could help guide resource allocation, preventing investment in protections against weather events that will *decrease* in severity in the future.

However common, statements like these are misleading since, as mentioned earlier, weather attribution studies *do not say anything about the risk of future* events, and are hence distinct from predictions or forecasts. Weather attribution only provides information about how *present-day risk* differs from risk in a natural environment (or past environments) (see Lusk 2017 for a more detailed argument on this point). Unless assumptions are made that the climate is static (which is known to be false), weather attribution does not have direct policy relevance where estimates of future risk are required. The perspective of probabilistic event attribution does not align with the goal of informing decision makers about future risks. Thus, we begin to see the value of a dual kind of analysis for adjudicating the value of methods of quantification for furthering our social goals.

Assessing the alignment between the methodology of event attribution and goals 2 and 3 requires deeper analysis. For goal 2 to be viable given the backward-looking character of extreme event attribution, the methodology would have to provide a fairly accurate view of how the present-day risk of certain kinds of extreme events has changed. This ties the social usefulness of event attribution to its reliability: unreliable attributions could lead to poor ethical decisions and maladaptive actions (Christidis et al. 2013). Similarly, if we want to separate victims of anthropogenically attributable extreme weather from those who suffered from naturally extreme weather for the purposes of recognition or compensation, as in goal 3, then probabilistic event attribution should be accurate and capable of attributing the breadth of event types that victims are likely to suffer from. Yet while many scientists (e.g., Thompson and Otto 2015) tout the reliability of event attribution, philosophical and scientific work rightfully questions whether confidence in event attribution results is justified.

While a full analysis of the accuracy of event attribution is beyond the scope of this chapter, examining two issues that arise when computer simulation models are used in attribution illustrates how problems of accuracy might result in a misalignment between attribution and its social goals. The first issue arises from the use of observational data to validate the reliability of the models used in attribution, and the second arises when scientists attempt to account for various kinds of uncertainties that emerge from the use of models. I examine each of these issues in turn.

First, scientists test the skill of their models against observational data for situations related to the ones of interest. Thus, they might deem a model adequate for attributing an extreme event if, in the region of investigation, the model reproduces the observed statistics for the relevant quantity (i.e., the distribution of temperature, precipitation, etc.). The results of a model are checked against the historical record: model adequacy is established when the distribution produced from an ensemble of many model-runs of a historical period matches the observed distribution for that period for the quantity of interest.

Tests of adequacy like this face two challenges. First, they rely on the reproduction of fairly robust statistical quantities that may not be sensitive to the weather extremes that scientists seek to attribute.[3] When scientists check their models, they want to see them reproduce the observed distribution across the entire range, including the tails where the extremes lie. However, scientists acknowledge that observations of extremes are too sparse to draw definitive conclusions regarding model adequacy for the extreme part of the distribution. Thus, they rely on other properties of the distribution to test their model. Stone and Allen (2005) write, "the only option is to look at more robust measures of the physical quantity, like its mean and variance, and assume that verification of our understanding of the physics and forcings relevant to these measures applies to its extremes as well" (313). The problem is that goodness of fit between model output and the nonextreme observations does not necessarily indicate fitness of the model for attributing extremes—particularly near record-breaking and never-before-seen events—that tend to lie in the tails of distributions. A model may perfectly capture the extant observations, yet its capacity to predict extreme events could still be questionable. There is thus reason to doubt that such models capture extreme events with the desired reliability.

The second challenge is demonstrating that models succeed in meeting validation criteria for the right reasons, that is, that they capture the mechanisms responsible for the event type in question. Capturing these mechanisms correctly underwrites the justification that the probabilities gathered in the natural scenario are realistic and is therefore crucial to demonstrating that estimates of FAR are reliable. Getting the right answer for the right reasons requires properly representing the relations between the mechanisms in the structure of the model. Scientists are aware that their models contain what is known as structural model error (SME) due to the nonrepresentation of relevant features, or inaccuracies in the

represented relations between features, and this knowledge of SME results in what are called "structural model uncertainties." What exactly
encompasses structural model uncertainty is still up for debate, but it is
construed by some as uncertainty about what would constitute a perfect
model and by others as uncertainty about what would constitute an adequate model structure for a particular purpose (see Parker 2010).

Event attribution studies typically employ only one model structure,
and thus the potential impact of different model structures is unac-
counted for in the (first-order) uncertainty associated with the results. It
is possible that two models, which both adequately reproduce observed
events, will differ in their predictions of events under the natural scenario.
This possibility raises doubt regarding the accuracy of model predictions
in the natural scenario, and is a source of what is called "second-order uncertainty" (uncertainty about uncertainty). Since these second-order uncertainties are potentially large, they undermine our confidence that we
have an adequate model structure in this case. Thus, these uncertainties
can be pernicious, and may undermine claims to the reliability of models
used in attribution.

Just how pernicious they are is a matter of significant philosophical and
mathematical debate. Recent results suggest that any SME might have a
significant effect on attribution results. Frigg et al. (2014) show that some
nonlinear dynamical models with SME are instable: small differences
in model structure can lead to very different trajectories for the system.
This is called the "hawkmoth effect," in homage to the well-known butterfly effect.[4] The notion of butterfly effect identifies a sensitivity to initial conditions: the same system begun from two close but nonidentical
initial conditions may have trajectories that diverge significantly. The
hawkmoth effect is similar but structural in nature: two very *similar but
different model* structures, started from the same initial conditions, can
result in very different trajectories. Hence, closeness in model structure
does not imply accurate or useful results. The hawkmoth effect could arise
in nonlinear dynamical climate models, and hence the hawkmoth effect
jeopardizes the accuracy of probabilistic predictions of climate states. The
kind of computer simulation models often used in event attribution are
of the sort that might be susceptible to the hawkmoth effect, and thus if
Frigg and his colleagues are right, such models would fail to be relevant for
decision-making.

The actual impact of the hawkmoth effect is not above contention,
however. Philosophers Winsberg and Goodwin (2016) place the burden of
proof back on Frigg and his colleagues, challenging them to demonstrate
that their conclusion has any impact on the use of nonlinear dynamical

models as actually employed in areas like climate science. They claim that the impact of the hawkmoth effect is determined by the presence of SME and instabilities in the model, but also a significant number of other factors: the time scale of the desired prediction, the means by which the results are obtained, and the question under study. This topic deserves more philosophical and scientific attention, but clearly the result of the debate could have significant impact on the way extreme event attribution is understood.

Examining the accuracy of one kind of event attribution, we can see that there are good reasons to think that at least this kind of event attribution is not yet reliable in the way that goals 2 and 3 require. Given the arguments above, event attribution could lead to maladaptive decisions, and therefore we have reason to think that its ability to support goals 2 and 3 are limited at best. Furthermore, it is widely admitted that even the limited reliability that probabilistic event attribution might enjoy applies to a small number of event types—predominantly extreme heat and cold—in limited geographic regions (National Academies of Sciences, Engineering, and Medicine 2016). Whether or not the capacities of extreme weather attribution will ever align with the social motivations for its use is an open question, but at the moment we have reason to believe that such alignment has yet to be achieved.

So far, we have considered how the epistemic aspects of the dominant forms of event attribution align with the social goals that certain proponents of event attribution have advanced. But we can also ask whether the values that scientists have employed in developing probabilistic event attribution align with or further the social ends to which event attribution may be put.[5]

As Lloyd and Oreskes (2018) have pointed out with regard to extreme events, the preference to avoid errors of a certain type have led to contentious debates about extreme attribution methodology that have a significant bearing on how event attribution studies can be used socially. Demonstrating how the chance of event occurrence has changed due to anthropogenic forcings is not the only way to attribute extreme events to climate change.

The so-called storyline method, developed more recently by Trenberth et al. (2015), attempts to clarify how particular events would have been different in a nonanthropogenically forced environment. Rather than look at how often a type of extreme event occurs, the storyline approach assumes the occurrence of a particular event and then attempts to assess how the atmospheric conditions that gave rise to that event would have been different in a "natural" climate. These assessments rely heavily on theoretical knowledge, particularly thermodynamical relations like the Clausius–Clapeyron relation, which specifies how much more moisture the atmosphere can hold per degree of warming. This approach therefore examines questions regarding why a particular event behaved as it did or what climate factors played a role in making the event so extreme—questions that are both conditioned on the occurrence of the event.

For example, Trenberth et al. (2015) analyzed a five-day extreme rainfall that occurred in Colorado in September 2013, which resulted in significant flooding in the area. Trenberth and colleagues reasoned that the event's development was significantly impacted by unusually high sea-surface temperatures thought to be the consequence of climate change. These temperatures were likely responsible, at least in part, for where and how the storm formed and for the significant amount of moisture it was able to carry and then release onto Colorado.

Proponents of risk-based probabilistic event attribution have vehemently objected to the storyline approach. Such proponents claim the storyline approach—given that it tends to examine thermodynamic features of storms at the expense of dynamical ones—would result in a failure of scientists' mission to serve society because almost every extreme event could be connected to climate change. As Allen (2011) notes, "blaming all weather events on climate change is not helpful either" (931). If so, this practice would jeopardize many of the social aims touted by proponents of the risk-based approach. Misattributing extreme events might also damage the credibility of scientists.

What Lloyd and Oreskes (2018) demonstrate is that much of the disagreement between advocates of the storyline approach and advocates of the risk-based approach rests on nonepistemic values that come to influence attribution science. For example, the two camps seem to disagree on what kind of information the public needs to be given; those advocating for a risk-based approach claim that "attribution analysis aiming to answer questions of *most relevance* to stakeholders should . . . follow a holistic [FAR] approach in order to prioritize an understanding of the changes of overall *risk* rather than the contribution of different factors" (Eden et al. 2016, 9; see also Otto et al. 2016). This reasoning—that the storyline approach fails to deliver the requisite information—leads to the view that permitting it would be a failure of scientific duty. But of course, the storyline advocates disagree.

Social values are similarly at the root of risk-based attribution proponents' claim that the storyline approach is not scientifically fit for social decision-making. Scientists operating under the risk-based paradigm value rigor and ensuring the number of false positives is small, even if this entails the risk of understating effects (type II error), in order to maintain scientific credibility. Scientists operating under the storyline approach have a different attitude to error: they are more comfortable with potentially overstating the effects of human actions (i.e., type I error). They value more highly tools that are likely to detect anthropogenic influences, even if it means false positives. The point to note is that the two methods of quantifying extreme weather are based on two different types of value commitments. In order to decide which should guide policy, one needs to know what goals a society is committed to.

If scientists' values may be carried with the quantified results they produce, there is a real danger that they may have an influence that goes unnoticed when used for social decision-making. As Porter points out, in many contexts numbers have an allure: once something is quantified, it seems natural to engage in judgment replacement. The ease with which numbers can replace judgment makes engaging in a coupled epistemic-social analysis all the more important.

Take, for example, goal 3: using risk-based extreme event attribution to identify the victims of climate change for ethical purposes like recognition and compensation (Thompson and Otto 2015). Risk-based methods, if they are accurate, quantify the change in frequency of a particular kind of weather event. In order to be compensated for damage incurred as a result of an extreme event, it needs to be demonstrated that the likelihood of the event had changed to some sufficient degree over time.

It would be easy then to put the burden on individuals wanting relief to demonstrate that the kind of event from which they suffered was anthropogenic in origin; that is, to require those harmed by extreme events to produce the right kind of numbers showing that the likelihood of the damaging event had increased. But of course, there are other ways to be victims of extreme weather for climate-change related reasons. The fossil fuels that power climate change, and the economies they run, have far-reaching consequences for land use, infrastructure, and culture that can significantly alter one's level of vulnerability. The same anthropogenic actions that cause climate change can lead to vulnerabilities that result in harm due to extreme weather events, regardless of whether the odds of those events have shifted. As Hulme et al. (2011) note, climate action is needed where vulnerabilities are high, not where storms are most attributable to climate change. Proposals that rely on risk-based analyses are likely to overlook victims of climate change who seemingly deserve recognition or compensation in the wake of extreme events but cannot produce the right kind of numbers (for more, see Lusk 2017). We need to articulate what kind of climate action should be achieved before we can say whether or not event attribution is sufficient to separate victims from nonvictims.

9.5 CONCLUSION

I have argued here that processes of quantification bestow on their results a certain perspective, despite attempts to achieve mechanical objectivity. This perspective is an important element that must be addressed when assessing the normative implications of using numbers in social policy decisions. As a minimal normative criterion of adequacy, I suggested that there must be an alignment between the perspective embedded in the process of quantification and the social goals that such quantification is intended to support. To demonstrate how one might go about assessing this alignment, I performed a coupled ethical-epistemic analysis of extreme weather attribution, showing that many of the ends touted by proponents of risk-based assessment seemed difficult to achieve given the current state of the science. The advantage of this kind of assessment is that it can make manifest which goals could be supported by a particular method of quantification, as well as what a method of quantification needs to look like in order to support particular social goals.

My employment of this kind of analysis is advanced only as a first step to answering normative questions about what kinds of numbers should be appealed to in social decision-making. One major drawback of the way I have deployed the ethical-epistemic analysis here is that certain dynamics of decision-making involving quantification are ignored. I have assumed that goals are static and that the development of methods of quantification does not impact those goals. Put another way, I have assumed that our goals do not change just because there is a convenient number available. The Original Critique suggests otherwise, namely that there is an appeal to numbers that does influence agendas. The Original Critique is certainly right on this point. But seeing the goals as relatively static does have at least one advantage, in that we realize we need not be seduced by numbers. To adopt certain numbers is to adopt a vantage point, and we should not do that uncritically when attempting to advance, or better specify, our social goals.

NOTES

1. Philosophical literature on measurement (see Tal 2013) and scientific inference from data (e.g., Bogen and Woodward 1988) embody this view.

2. Here I'm taking inspiration from the "adequacy for purpose" account of scientific modeling (see Parker 2009) which, in short, claims that the standard for scientific model evaluation should be adequacy for a specified use, rather than a true image of the world.

3. It is worth noting that the difficulty here in demonstrating model adequacy is unique to this practice and stems directly from its focus on extreme events. Many other applications of climate modeling do not encounter this problem because they deal with more robust quantities, like yearly global mean temperature.

4. Erica Thompson, a colleague of Roman Frigg and his coauthor Leonard Smith, originally coined the term "hawkmoth effect."

5. Whether social values (e.g., value of human life, religious values, personal preferences) should or inevitably do influence science when producing quantitative results that are socially relevant, are topics of lively debate in philosophy of science (see Douglas 2009), especially when it comes to climate science (see Betz 2013; John 2015; Steel 2016; Winsberg 2012). The emerging consensus seems to be that social values are unavoidable in this kind of work, though there are ways to handle social values such that they play a legitimate role in scientific reasoning.

CHAPTER 10

The Purposes and Provisioning of Higher Education

Can Economics and Humanities
Perspectives Be Reconciled?

Aashish Mehta and Christopher Newfield

I wonder whether you get information coming into college that says you know, this course of study will lead to this kind of jobs and there's a lot of opening here as opposed to—as you said, English—and as an English major I can say this . . . as an English major your options are uh, you better go to graduate school, all right? And find a job from there.

—Mitt Romney, April 2012

I promise you, folks can make a lot more, potentially, with skilled manufacturing or the trades than they might with an art history degree.

—Barack Obama, January 2014

I made fun of philosophy 3 years ago but then I was challenged to study it, so I started reading the stoics. I've changed my view on philosophy. But not on welders. We need both! Vocational training for workers & philosophers to make sense of the world.

—Marco Rubio (@marcorubio), Twitter, March 28, 2018

10.1. INTRODUCTION

Scholars in the humanities disciplines and scholars in economics have said little to each other over the years about the point of graduating from university or how society should pay for it. This is not surprising. The lack of scholarly communication across this particular disciplinary boundary extends to many other areas of public policy. Economists prefer parsimonious explanations of human behavior, axiomatic answers to normative questions, and decision-making based on quantitative evidence. These intellectual instincts are about as far removed as one can get from humanities scholars' interest in individual variety and psychological depth, insistence on historically and culturally contextualized approaches to the normative, and comfort with qualitative evidence and arguments.

This chapter reflects the efforts of two public university professors—a humanities scholar and an economist—to bridge this divide. Our informal conversations about the state of higher education led us to collaborate on the international research project of which this volume is one result. We found that we mostly agreed on the public benefits of higher education and on the limits of quantified assessments of its value. We concurred that the accepted purposes of university education seem to be shifting further and further from those endorsed by the humanities. We also generally agreed that private funding models would underfund universities and exclude much of the population that would most benefit from university attendance. And yet we were from disciplines that are often publicly perceived as being at odds on these questions.

The lack of communication between our intellectual traditions has become increasingly consequential. Policy has not waited on some kind of interdisciplinary accord. Higher education institutions are increasingly managed according to principles that purport to reflect economic reasoning. The privately borne costs of university attendance have increased dramatically, and this has had profound implications for how many go to college, who goes to which colleges, and who majors in what. State funding of public universities continues to decline (Newfield 2016). Racial and income gaps in access to better-funded universities remain high (Carnevale and Strohl 2013; Chetty et al. 2020). Unprepared to discuss the monetary value of humanities education, the departments offering it have also experienced declining enrollments ("Bachelor's Degrees in the Humanities" 2021).

The accepted purposes of university education therefore seem still to be shifting further and further from those endorsed by the humanities, with little communication between economics and the humanities disciplines about how to theorize this shift or what to do about it. This chapter describes the results of our efforts to grapple with this issue. These efforts consisted of shared readings from both traditions, extensive discussions, and a detailed content analysis of dozens of economics textbooks and journal articles. This process has, we believe, enabled us to arrive at substantive agreement on several points.

We find that economists' understanding of the benefits of education, as reflected in human capital theory,[1] is potentially compatible with the views of most schools of humanities thought. However, we argue that several features of the economic approach ensure that not all of what humanities scholars think higher education is good for makes it into economic research, and also that not all of what economists think higher education is good for makes it into public discourse. What get left out, disproportionately, are the benefits it delivers that the humanities remind us are important and that may well be the relative strength of humanities education.

The reasons for this are many but quantification does loom large. The benefits of education of interest to humanities scholars take the form of changes within the person being educated—in how they think and in what they value and can understand. These changes themselves are difficult enough to define, much less measure, so economists have limited themselves to measuring differences in outcomes that they interpret to be projections of this personal growth onto the observable plane. The benefits they are able to document leave out many of the benefits of higher education laid out by centuries of humanities thought. We argue that this has contributed to policies that fail to promote the delivery of these benefits.

The arguments are nuanced, covering more than 200 years of intellectual history in the humanities and over sixty years in economics, and a fairly large set of conceptual ideas. In the interests of readability, we therefore telescope our claims and get on with making them. We are motivated in part by this question: What if, regarding the purposes of higher education, humanities scholars and economists are not philosophically opposed so much as separated by different relations to the numerical?

We begin, in section 10.2, with what we will refer to as "the humanities view," which we stylize in the interests of space. Humanities scholars have not treated the variety of possible university financing arrangements in as great detail as economists have. However, they have put much more energy into thinking about what university education is for. They argue that higher education effects cognitive and psychological changes, and that these produce benefits that are heavily nonmonetary or accrue to society at large—not just to the person being educated. These benefits include self-formation, in conjunction with knowledge formation, and a broad-based capacity for social critique. Scholars in the humanities also describe in detail how the *process* of education is integral to these objectives—a consideration that is surely relevant to the determination of how universities should be organized and financed. Finally, humanities scholars are compelled, by their conception of what education is for, to reject neoclassical economists' tendency to treat equality of opportunity as an axiomatic ideal that might be considered separately from economic efficiency: if higher education is partly about building society, then it is not possible to discuss the total benefits and costs of education without taking the present state of society—and therefore, the historical distribution of access to education—into account.

TABLE 10.1. An economist's taxonomy of the benefits of higher education, with examples

	Private benefits	External benefits
Pecuniary benefits	• Increased personal productivity • Higher expected wages • Higher-earning partners	• Increased productivity of coworkers • Increased earnings of employers
Nonpecuniary benefits	• Increased enjoyment of culture • Better emotional relationships	• More vibrant democracy • Greater environmental awareness

Section 10.3 lays out most education economists' understanding of the value of higher education and the pros and cons of different ways of financing it. Economists categorize the observable benefits of education using a standard 2 × 2 taxonomy, illustrated with examples in table 10.1. Any observable benefit of education can be private (accruing to the person being educated) or external (accruing to others); and also pecuniary (it consists of *financial* rewards) or nonpecuniary (the rewards are nonfinancial). A key result is that if the observable benefits of higher education are all private and pecuniary, and if credit markets function seamlessly, a privately financed higher education system will be efficient. The converse holds as well—if a large share of the observable benefits is external and nonpecuniary, as humanities scholars insist, then a privately financed system will generate significant inefficiencies. It will also, we argue, yield results that are inconsistent with notions of fairness that legitimate efforts to establish universities in the first place.

In section 10.4 we show how human capital theory has been mobilized by economists over time to demonstrate that they are indeed aware of and committed to all four categories of benefits in table 10.1. This means that most of the benefits articulated in the humanities are indeed compatible, at least in theory, with economic approaches to designing university funding arrangements. We also make the case, using economics textbooks and journal articles, that the nonpecuniary and external benefits of education (emphasized in the humanities) receive less sustained attention in economic writings than do private, pecuniary benefits. We show that this is because of a lack of data acceptable to economists and the particular difficulties of causal identification that arise with respect to external benefits. These arguments apply especially well to the types of nonpecuniary and external benefits claimed by the humanities, which are extremely hard to quantify. We argue that this could help explain the current social narrative, put forward by policymakers, in which the primary role of education is to train workers, and produce a form of social equity reduced to more equal access to plum jobs via university diplomas.

Section 10.5 considers and acknowledges the limits of our argument, discusses points of consensus on how to finance higher education, and concludes.

10.2. THE BENEFITS OF HIGHER EDUCATION: PERSPECTIVES FROM THE HUMANITIES

In this section we offer a very restricted list of key effects of studying humanities fields in the university. We limit ourselves to an American lineage with a key German and British precursor. We do this for two reasons. First, this tradition formed the US research universities that housed academic economics in its postwar era of American hegemony. Second, this tradition combines individual and external *non*pecuniary benefits in a distinctive and useful way. For example, it emphasizes personal development, citizenship, and later, racial equity. None of these are pecuniary benefits but the first is individual, the second is external, and the third is a combination of the two. Given restricted space, we are unable to analyze a representative range of views, and sample only those that directly address the purposes of the university. This is the primary reason why the humanities fields are represented in this section mainly by critical theory and ethnic studies.

The humanities view of university effects has three general features that have persisted through the past 250 years. These fields cast the effects as (1) overwhelmingly nonpecuniary. Humanities scholars (2) ignore or downplay the distinction between *private* and *external* benefits that has been important in economics. In addition, they (3) generally favor inclusion of students without regard to private income, within the usually severe racial and gender restrictions of a given place and time. Yet these fields have never developed a financing model for the support of universities' complex and labor-intensive nonpecuniary effects.

10.2.1 Bildung, *the Core Nonpecuniary Benefit*

The humanities' foundational theory of the modern research university was developed in the late 1790s and early 1800s in Prussia. It was a direct response to one of the West's periodic crises of the boring professor.

When professors become so boring that they are seen to undermine the value of higher education, political and business powers—or, in this case, the king of Prussia—plot to throw the baby out with the bathwater: they try to get rid of professors by getting rid of the university, usually in order to replace it with a vocational school that will directly cater to the state and the economy. This was true in Berlin in the first decade of the 1800s.

The boring professor is always boring for the same reason: he has let his teaching methods fall behind technology, which has rendered his methods obsolete. In 1800, the relevant learning technology was the printed book. The medieval university had developed the lecture so that students could copy the professor's reading of the one local version of a hand-created book that was likely to exist. Before Gutenberg, students had to listen to a professor read because the professor possessed the text. In the post-Gutenberg world of multiple copies, why should the professor read and the students transcribe?

There was the further matter of the possible irrelevance of Kant's "lower faculties," centered on philosophy. German universities suffered soft enrollments, and the job-oriented professional courses were doing better than the others (Menand et al. 2016).[2] In a double move that is familiar today, reform required replacing professorial lectures with new technology and what we now call the liberal arts with more vocational majors.

This was the situation that prompted the renaissance of Prussian university studies that has brought us a theory of the research university that we continue to use (e.g., Wellmon 2015). We're going to call this diverse body of complex argumentation the "humanities' theory of the university," and summarize some key elements for purposes of this chapter's comparison with economic views of the university's mission. In addition to being simplified and idealized, this account will also line up potted ideas from the 1800s, 1880s, 1930s, 1960s, and 2010s. We are suppressing historical specificity and diversity in order to identify a broadly humanistic understanding of the transformative effects of higher learning that has formulated alternatives to instrumentalist definitions of the university for at least 200 years.

To avoid the dismembering of the university as a medieval relic, the key university theorists of the 1790s and early 1800s, including Herder, Fichte, Humboldt, Schelling, Schiller, and Schleiermacher, had to explain why the university was neither a glorified high school nor an excessively unspecialized professional school. The most basic feature of the true university was that it served only one thing—the progress of knowledge. It did not ever serve any specific interest, be it private pecuniary interests or interests of state; worldly goals would distort the pursuit of truth. The distinguishing feature of the true university was that it never strayed from putting the search for and dissemination of *new* knowledge ahead of all instrumental interests, even its own. In this quest, teaching and research were the same thing, just involving different kinds of people.[3] Though many of the university's faculties had a professional orientation, and though it helped graduates and the society pursue pecuniary goals, the university's *distinctive* mission was to serve *non*pecuniary interests. Knowledge had to be its overwhelming and defining priority or it would not be a university.

But how did students and their professors learn to put the pursuit of knowledge ahead of the interests of themselves and others? The university taught not just subject content (as in high school) but also the whole system of human knowledge and the procedures for continuously expanding it. One could become a university student in the true sense only if one grasped that learning was (1) a *process* that was (2) unending—mastery would never take the form of a final understanding—and (3) an active personal engagement inseparable from self-creation.

This last point was particularly fundamental. The university taught the student how expanding knowledge was the same process through which they formed their own identity and enlarged self-knowledge. Learning was *Bildung*, or self-formation.[4] Creating knowledge was intertwined with cultivating the self. The university taught students how to spend their life "converting as wide a range of experience as possible into conscious thought"—to contribute to knowledge by grasping identity-based experience.[5]

What were the conditions of university study, which was a form of life, an existential orientation, a mode in which one's own being unfolded in relation to that of others, day after day, via a conscious embrace of being as a form of knowledge?

A primary condition was financial freedom: the university should sustain itself financially without its members needing to devote themselves to its funding. Students with the proper achievements but without family funding should receive free educations. Professors should not need to take second or third jobs, or steer their research toward the ideas and interests of the local gentry. Wage earners naturally resented people who seemed not to need to work for a living, but the Prussian theorists didn't flinch. If anything, they doubled down. Cultivating knowledge and the self required complete academic freedom, which Schleiermacher defined this way:

> To live on the streets like the Ancients, to fill them with music and singing like people from the Mediterranean, to feast like the rich, as long as

money lasts, and then to despise all comforts of life like ancient Cynics, completely to neglect one's clothing or to dress up fashionably . . . that is academic Freedom. (Cited in Schmidt 2013, 172)

The university's practices of knowledge were rooted in freedom from both convention and pecuniary constraint. The Prussian theorists took the risk of sounding irresponsible in order to insist on knowledge's absolute need for freedom from want. Humboldt, for example, suggested to the king that he grant the proposed University of Berlin an endowment that would allow it political and economic independence from the state. Neither teaching nor research should be limited by what we would now call pecuniary factors. Even more fundamentally, the university's contribution to human development depended, for these theorists, on the repudiation of the economic view of society. This is a major reason why their diverse intellectual descendants did not develop an economic theory of funding university education under conditions of scarcity.

Society was supposed to pay for universities to spare them pecuniary restrictions so students and faculty could focus their minds on nonpecuniary issues of truth, which was in turn linked to the historical unfolding of humanity's destiny. Did this mean that these conditions of intellectual freedom should be reserved for a small, brilliant elite? The general answer was no. The university sought not the private gain of individuals but the *collective* progress of the entire public. This does not mean that the Prussian theorists were imagining universal university attendance. The universities were not inclusive, but their effects were to pervade all of society.

Thus the Prussian theorists avoided the standard economic distinction between private and external benefits. In Fichte's (1988) words, "the demand that every person ought to cultivate all of his talents equally contains at the same time the demand that all of the various rational beings ought to be cultivated equally'" (166). Not everyone had to attend university, but enlightenment and emancipation required the full cultivation of every member of society. Society therefore needed a general Bildung that the university modeled and theorized. The knowledge that the university furnished for this project was not merely contemplative knowledge. It was not a stock to be possessed but an ongoing process that led to action, and which, because it was active, joined "private" and "external" development as two sides of the same general cultivation.[6]

By 1810, the modern university had a body of humanities-based theory that identified the individual intellectual capabilities that were in turn the necessary conditions of historical progress. The university's deep value was fundamentally nonpecuniary and simultaneously private and external.

Though its most powerful sectors were those "higher faculties" that provided direct expertise to the state (theology, law, and medicine), the "lower" philosophical and humanistic faculties judged the validity of professional thought and action, and thus guarded the collective destiny.

These ideas appeared in the English-speaking world in part through the travel of US and UK intellectuals to German universities. In university theory, the pivotal Anglophone was John Henry Newman, whose *Discourses on the Scope and Nature of University Education* (1852) conceptualized the university as the unique place of the integration of all knowledges that could not be reduced to their use. Newman famously defined liberal knowledge via Aristotle as tending toward "enjoyment." By this he meant knowledge whose possession was its own purpose: though it may well have a practical use, its value did not depend on its utility but on its status as (true) knowledge (Newfield 2003, 55–56). Liberal and practical knowledge coexisted, and yet must not be reduced one to the other. "Knowledge for its own sake" was not impractical or merely "academic" knowledge. It was knowledge that existed in itself, and did not require legitimation through the later arrival of pecuniary value. This is still a difficult point for Anglo-American policymakers to grasp.

Crucially, Newman defined the nonpecuniary value of knowledge as the *precondition* of professional and other practical forms of knowledge (those which had political and economic as well as intellectual value). A university is an "assemblage of learned men, zealous for their own sciences, and rivals of each other," who "learn to respect, to consult, to aid each other" so that they may each "apprehen[d] the great outlines of knowledge, the principles on which it rests, the scale of its parts." This is "liberal" education—that which allows multiple disciplines to appear in their relations to each other. The university enabled this seeing of knowledge in its interrelations by bringing the sciences, arts, humanities, and social fields together. It could then become a "habit of mind . . . which lasts throughout life." The university must "bundle" knowledge. Unbundling it according to financial calculations would deprive it of its distinctive features.

Nothing in Newman's view of liberal education blocks the teaching of practical subjects like law or medicine. These are clearly taught in universities, but they are "not the end of a University course" of study (Newman 1996, 77–78, 118).[7] Liberal knowledge coexists with and may lead to pecuniary gain, and yet, to repeat, it is knowledge that does not *aim* at pecuniary gain and is not controlled by pecuniary logic in any way. The university's core function is to support "the most important non-market public good, [which] is knowledge" (Marginson 2016, 88).

10.2.2 Equal Opportunity

In the midst of the American Civil War, Congress passed the Morrill Land-Grant Act, which gave states up to 30,000 acres to establish public colleges for their local populations. The Morrill Act was part of the nationalist expansion and settler colonization of the North American continent. It also reflected the dual aspect of the Prussian research university aspiration, which was to organize a national population while at the same time establishing the capacity for intellectual emancipation. In the United States, the national population was organized in part through the killing or removal of the Native population, and public universities aided a process of implanting the settlers who took the land. The author of the act, Justin Smith Morrill, wanted land-grant institutions to admit freed slaves to colleges after the union was restored: "They are members of the American family, and their advancement concerns us all. While swiftly forgetting all they ever knew as slaves, shall they have no opportunity to learn anything as freemen?" (Fretz 2008). Admission of former slaves did not happen. When a second Land-Grant Act (1890) addressed African American higher education, it did so by setting up a separate and unequal system of vocationally oriented colleges that reinforced Jim Crow segregation.

Within its strict racial demarcations, the Morrill Act of 1862 is often read as establishing practical colleges for local populations. It also shared Fichte's rejection of the contrast between thinking and doing and Newman's hostility to polarizing the liberal and the practical. The land-grant colleges were,

> without excluding other scientific and classical studies and including military tactic, to teach such branches of learning as are related to agriculture and the mechanic arts, in such manner as the legislatures of the States may respectively prescribe, in order to promote the liberal and practical education of the industrial classes in the several pursuits and professions in life. (Fichte 1988, 7)

The principle of higher learning for all was embodied in the "liberal *and* practical" mixture, offered to the "industrial classes" (emphasis added).

One of the University of Michigan's important early presidents, James Angell, spelled out the meaning of public higher learning in a commencement address delivered at the end of June 1879. Angell's thesis was in his title: *The Higher Education: A Plea for Making It Accessible to All*. That last phrase meant making the university "accessible to the poor as well as to the rich." It also meant, he said explicitly, including Black students as well as white. The body of his address gives five reasons why excluding most of the population from "the higher education" would be, in his phrase, "in the highest degree unwise" (6). The closing section makes a second pointed claim: perhaps Harvard and Yale can exist on private funding, but the great universities of the developing West cannot. These Western universities absolutely require public funds for the sake of general access to higher learning. Finally, he insisted that what would be nearly free college needed to provide "liberal education" to everyone able to receive it. Liberal education would not go to elites while the poor and Black got only vocational training.[8]

But whatever the vision of racial equality proposed on formal occasions by university leaders like Angell, it was honored in the breach. By the time W. E. B. Du Bois published *The Souls of Black Folk* (1903), the dominant paradigm was "common school and industrial training" for African Americans and Native Americans—at best. This was the system that Booker T. Washington had to fight for, since even common schools depended on "teachers trained in Negro colleges, or trained by their graduates," whose existence was always threatened via the suppression or marginalization of their institutions (23). It was this system that DuBois harshly criticized as inadequate and self-contradictory.

The United States operated a version of racial capitalism that was at the same time a racial democracy, a rigid color hierarchy. And yet some university advocates successfully articulated general and inclusive social development as a fundamental public good—one that was largely external to individual pecuniary provision and that required public funding.

In this context, it is unhistorical and inconsistent with universities' founding principles for economists to treat equality of opportunity as a purely normative concern that they can set aside to perform a valid cost-benefit analysis.

10.2.3 Democratization of the Intellect

In the United States, the 1940s brought major shifts in federal higher education theory and practice that led to massive increases in university enrollments: they rose 78 percent in the 1940s, 31 percent in the 1950s, and 120 percent in the 1960s. The Serviceman's Readjustment Act of 1944 created a series of publicly funded benefits for war veterans that paid university tuition and living expenses. The legislation helped establish the principle of what more recently has been called "free college"—public colleges and universities that should be open to all who are motivated to benefit from them, regardless of income or even academic record.

Two 1940s reports declared a democratic moment in higher learning. *General Education in a Free Society* (Harvard University Committee and James Bryant Conant 1945, known as the Harvard Redbook) called for "general education—a liberal education—not for the relatively few but for a multitude." *Higher Education for Democracy* (President's Commission on Higher Education 1947, known as the Truman Report) asserted that higher education must eschew its traditional elitism to "become the means by which every citizen, youth, and adult is enabled and encouraged to carry his education, formal and informal, as far as his native capacities permit" (101). The historian Michael Meranze (2015) has noted that both reports, though national security documents, put the humanities at the center of general education for a crucial reason. Social and political knowledge were deemed as important as technical knowledge to the functioning democratic societies (1313–15).

The postwar period evolved an understanding of the nonpecuniary and external civic benefits of widely accessible higher education. The period's public colleges and universities were basically free but also predominantly white—97 percent white in 1940, and still 80 percent white in 1995. Full access *and* intellectual self-determination were put on the agenda by the Black civil rights movement and Black Studies programs and departments. Together with feminist studies and other ethnic studies initiatives, Black Studies contested the (white) nationalistic understanding of university-based general development.

Black Studies was a joint production of the humanities and social sciences, and one of its founding motivations was that the US university did not sponsor objective scientific knowledge about society and culture so much as it sponsored variations of scientific racism that justified racial exclusion and stratification. Kant, Hegel, and other foundational theorists of the enlightenment university were themselves racists, and the specifics of their vision of human transformation could not be separated from what we would now call their white supremacism (e.g., Boxill 2017). Later in the nineteenth century, key humanities disciplines formed around Eurocentric canons of masterworks (English, Art History) and ethnocentric understandings of what was worth studying, how to study it, and the personal and external effects of studying. The Black Studies critique did not spare the natural sciences, which had been directly involved in developing empirical "proof" of the inferiority of all non-European races and ethnicities, as well as measures of supposedly innate aptitude that had ensured their underrepresentation in higher education.

Put another way, Black Studies saw institutional and intellectual exclusions as two sides of the same coin. It was thus consistently engaged in a critique of academic knowledge as it emerged from the structural biases present in all disciplines. While repudiating the tradition's racism, Black Studies also advanced the virtue of the Kantian "lower faculties," which was to judge the philosophical legitimacy and validity of the "higher faculties" that served government and business. The higher faculties resisted this, as did political science and other social sciences. Small numbers of scholars in such disciplines participated in the critique of the basic "terms of order" of the disciplines (Robinson 2016). Most did not. As the political scientist Charles P. Henry (2017) remembered, "It was in Black Studies, not mainstream political science, that [one] found a challenge to the dominant paradigm" (158).

The main line of defense against Black Studies was to call it politicized knowledge, in alleged contrast to objective knowledge.[9] This charge did not stop the early figures in Black Studies from supplementing their critique of existing knowledge with a new epistemology grounded in Black (as well as feminist and, later, queer) standpoints. Valid knowledge emerged not only from rigorous methodology but from the actual experiences of communities that had been pushed to one side in imperial and/or racist Western cultural systems. Standpoint epistemology in Black Studies, feminism, and other fields, is compatible with the understanding of Herder and Hegel of the emergence of truth from rigorously analyzed experience. Standpoint epistemology required that Black Studies be established in independent departments, rather than in programs where its methods and subjects could be controlled by those of the mainstream disciplines that had caused the problems Black Studies sought to correct.

Black Studies thus made a parallel set of demands for Black education both inside and outside the university. On the one hand, its scholars sought to decolonize knowledge production and "provincialize" the European contribution (Chakrabarty 2000). On the other hand, they called for autonomous educational institutions in Black communities. The autonomy of these institutions would enable them to produce knowledge grounded in the presumption of the equality of African-descended cultures with those from Europe. This in turn would enable Bildung that emerges from the possibility of full self-respect—which requires, as a first step, that one's cultural and intellectual heritages be respected on equal terms and not be seen as inferior to those of the dominant society (Hammond 2014). The same was to happen within the university, where the knowledges produced by Black Studies were to be equally valued, on the whole and accepting individual variation, as those produced by Political Science or Sociology.

Black Studies sought both epistemic parity in academic knowledge and ethnocultural equality in the society as a whole. In so doing, it demanded that the university finally truly support the free development of members of oppressed groups. Access to universities, institutional support for social justice, and the pursuit of new and diverse knowledges all meant that the university needed to be willing and able to pursue nonpecuniary goals.

10.2.4 Supporting Critique and the Analysis of Identity

The post-1960s humanities fields flourished in large part because, for decades, they were not subject to cost-benefit analyses asking them to demonstrate a direct monetary return on investment. When that question did arise (as it did later in the 1960s and 1970s, and again with increasing persistence after the financial crisis of 2008), the humanities fields were largely unprepared, having spent the previous two centuries focused on a nonpecuniary model of the inputs and effects of knowledge creation in universities.

Vulnerability notwithstanding, the work on nonpecuniary benefits produced major intellectual and social effects. The reciprocal development of ethnic studies and civil rights movements has changed the racial expectations of American society while delegitimizing Eurocentric knowledge. A case in point is Ta-Nehisi Coates's description of his youthful long-term response to Saul Bellow's rhetorical question "Who is the Tolstoy of the Zulus?," which was to reject the premise of the question and then study, meaning study everywhere in society but also study with particular depth and structure in the university (Coates 2015).[10]

The rise of critical theory in US universities was coextensive with new analyses of identity that are better called *the critique of identity* than the more familiar term, *identity politics.* An exemplary case is the work of the philosopher Judith Butler, whose breakthrough text, *Gender Trouble* (1990), critiqued normative understandings of the essential stability of gender. This work reflected a large collective intellectual and practical effort, much of it inside the university. The field converted many people to the view that gender is mobile and nonbinary.

In a 2009 essay, "Critique, Dissent, Disciplinarity," Butler clarified the importance of the university in supporting the work of critique, which in turn consolidates and justifies dissent. There are "two dimensions of [Foucault's] notion of critique," she wrote, "and they are interrelated: on the one hand, it is a way of refusing subordination to an established authority; on the other hand, it is an obligation to produce or elaborate a self" (787).

This first element is fundamentally important. Like their predecessors, contemporary humanities theorists have insisted that the obligation to submit to existing authority generally damages or blocks the pursuit of knowledge. Autonomy is constituted in the process of freely giving or withholding consent, and no true research or teaching can take place without it. In Butler's terms,

> what should be preserved as a value of the university is precisely that operation of critique that asks by what right and through what means certain doxa become accepted as necessary and right and by what right and through what means certain government commands or, indeed, policies are accepted as the precritical doxa of the university. (2009, 783)

The doxa to which literary studies and critical theory paid special attention were identity categories that claimed natural status and authority on that basis (cis-male gender, heterosexuality, European culture).

The second aspect also descends from Herder and Kant, as well as from Arendt and Derrida: the creation of adequate as opposed to oppressive knowledge is always connected to self-reflection about the categories and processes of knowledge production *and* in relation to self-creation.

Both features have subjected the humanities, particularly the "studies" disciplines (feminist, ethnic, cultural), to fifty years of political attacks.[11] Opponents of critical theory were also displeased when these features worked together, and critique asserted its authority over established political and social power on the basis of an intersubjective and collaborative intellectual *process* rather than an epistemological *ground*. A crucial theme of this wing of the humanities at this period has been the rejection of foundationalist epistemology (asserting objective grounds for knowledge that are independent of the perspective of the perceiving subject) and its replacement by nonfoundationalist procedures governed by evolving professional practices as protected by academic freedom from external pressures.[12] Truth was never fixed or final. As the literary critic Michael Wood put it: "The fine point of deconstruction is not that we can't decide but that we have to keep deciding" (Wood 2015).[13]

The complexities of these issues go beyond our scope; suffice it to say here that these forms of interpretation and critique are hard to do. *Learning how to keep deciding* was one of a remarkable range of nonpecuniary effects that the humanities offered through the university. Others were the capacity to dissent from established authority, to cultivate an autonomous self, to operate without the need for artificial epistemological grounds, to undermine a dominant culture's claim to intrinsic superiority, to critique the normative identities and relations that claimed natural authority, and to possess the collaborative powers that can democratize institutions.

It is notable that thinkers who claim a "humanistic baseline" often view education as furthering "the development of an individual qua human being—namely, a creature whose flourishing entails the development of a range of valuable cognitive, affective, and intersubjective capacities" (Allen 2016, 14). This writer, the philosopher Danielle Allen, emphasizes the classic connection between Bildung and full social agency, which, if achieved, would mean intellectual power over political power as such.

In sum, the branches of the humanities most engaged with the function of universities have focused consistently on articulating and developing the nonpecuniary effects of higher education, both private and external. Over time, these nonpecuniary effects have ranged from procedures of individual development to analysis of the conditions of racial justice to the study and legitimation of sex-gender nonfixity. These diverse topics have in common the conjunction of knowledge and identity in a continuously unfolding and structured reflection on experience—and their status as effects that cannot be measured in monetary terms.

At the same time, humanities disciplines have neglected pecuniary dimensions, in terms of its effects for both individuals and the economy, and of the funding structures required to support intellectual evolutions of all kinds. Modern economics enters at this point.

10.3. HUMAN CAPITAL THEORY AND THE FINANCING OF HIGHER EDUCATION

Human capital theory (HCT) was developed in the late 1950s and 1960s, in response to the observations that richer, more productive countries are more educated, that employers also find more educated workers more productive and express this in wages, and that while employers frequently complain about shortages of trained workers, people struggle to access education. This then prompted the question: What induces investments in human capabilities? HCT attempted to provide transparent, testable responses to this and related questions.

What follows is a discussion of how HCT views the world, with an emphasis on explicating the normative assumptions it invokes. We will clarify the implications of these normative assumptions as they are introduced and return to the theory's (mis)application at the end.

Becker's seminal 1964 treatise on the subject defines investments in human capital as "activities that influence future monetary and psychic income by increasing the resources in people" (Becker 1993, 11). There is a massive body of theoretical and empirical work on investments in people,[14] and the term "human capital theory" is used in many ways in this literature.[15] In this chapter, we use the term to reference the set of theories that adopt ideas from neoclassical (marginalist) economic theory to explain observable behaviors with respect to investments in people.

In keeping with this tradition, HCT lays out the market and nonmarket forces that would push private actors to make investments in human capital. It lays out the circumstances under which these forces would lead to inefficient investment patterns. The private actors whose behaviors are modeled include students, trainees, and workers—the individuals in whom these resources are embedded; their families—who shoulder some of the costs; employers—whose cooperation is often required to make these investments feasible and whose willingness to pay for these resources renders investing in them financially attractive; and for-profit trainers and educators, whose efforts contribute to the production and transmission of the knowledge. HCT explores the incentives of these actors to invest in education (and health, although we set that aside for this chapter) and prescribes public actions, undertaken by governments, to correct any failure by private players to promote efficient investments in people.

Economists attend to a much narrower set of normative objectives than humanities scholars. This permits a sharp, axiomatic treatment to each normative objective while putting some public concerns out of analytical range. Efficiency is the central normative objective considered by HCT, so it is essential to understand exactly what it means. By definition, an investment in people is efficient whenever it brings benefits to society that are larger than the cost of that investment to society. The benefits to society are the sum of private and external benefits, while social costs are the sum of private and external costs. Inefficiencies arise when investments whose social benefits exceed their costs are not undertaken or when investments whose social costs exceed their benefits are undertaken. Inefficiencies are to be avoided so that resources can be reallocated to more productive ends.

Three points about this definition bear clarification. First, as explained in table 10.1 and section 10.1, it concerns all social benefits, not just private benefits and pecuniary benefits. The 2 × 2 classification of benefits as private/external and pecuniary/nonpecuniary is central to HCT, which demonstrates that the efficiency implications of different approaches to education finance depend upon the types of benefits the education delivers.

The second point to be made about the definition of efficiency is that benefits and costs, which must be measured in dollars (or whatever currency), reflect the circumstances of the person in whom an investment is being considered, and those of their family. Specifically, benefits and costs in economics are measured by what people are willing to pay for something, which is in turn a reflection of their wealth. This has its virtues, so long as we are comparing benefits across persons of similar means. In the present context, it ensures that it will be efficient to grant the poetry aficionado a seat in a poetry seminar over their philistine sibling: the aficionado would feel more enriched by the experience and would be willing to say it with cash. On the other hand, the measure fails to distinguish the philistine from the merely less affording. If the lawyer's philistine daughter, by virtue of the resources at her disposal, is willing to pay more to attend a poetry seminar than is the maid's more culturally inclined daughter, then the benefits of attending the seminar are considered greater for the lawyer's child.[16] Thus, economics' use of a particular quantitative notion of value, selected to allow investments in the education of different people to be treated as commensurable, implicitly ignores variations in people's ability to pay for education.

The third point is that in assessing economic efficiency, economists do not count greater equality in the access to education as a benefit. Rather, while many economists do study it, they treat it as an independent goal. Indeed, efficiency is not supposed to safeguard any notion of fairness. It is defined in this way in an attempt to separate the more technical question of how to make the most of societies' resources from the more political questions about who should benefit. An education system that perfectly replicated existing social or racial hierarchies could be efficient so long as it did not detract from productive possibilities. As noted in sections 10.2–10.4, humanities scholars reject this separation of objectives. For surely educating students from working-class or historically disenfranchised backgrounds benefits society as a whole, in addition to improving life for the student themselves. And, if the costs and benefits of educating different people depend upon past inequities, the separability of equity and efficiency in the present borders on subterfuge.

We emphasize that table 10.1 is a taxonomy of the possible externally observable benefits of higher education. It is not a taxonomy of types of education. Any educational activity could yield multiple benefits, and these benefits could fall into multiple boxes. For example, a course in civil engineering will typically increase the productivity of the worker herself (a private, pecuniary benefit), the productivity of her colleagues (an external, pecuniary benefit), her enjoyment of an elegant solution to a problem (a private nonpecuniary benefit), and her commitment to building infrastructure that is ecologically sound (an external nonpecuniary benefit). While the taxonomy is very helpful for clarifying the likely effects of different financing arrangements on individuals' educational investments *in theory* (demonstrated below), its usefulness in practice requires understanding empirically this one-to-many mapping of education to benefits.

Even if such an empirical mapping could be made, it would not sit well with humanities scholars' conceptions of what education does. For, as noted in section 10.2, the interconnection of self-development and knowledge formation is the core end, and to the extent that the observable effects of achieving that end matters, the benefits that result are likely to be both private and external, as well as pecuniary and nonpecuniary.

Economists study policy questions by drawing a bright line between normative and positive analyses. Efficiency is a normative criterion (as is equality of opportunity). It is one way of deciding what *should* happen, in pursuit of society's best interests. Analysis of what *actually tends* to happen (positive analysis) is a discrete activity to be undertaken separately from answering normative questions. In HCT, the private/external and pecuniary/nonpecuniary distinctions do not matter at all for determining whether an investment in human capital should be made—efficiency dictates only that those investments in human capital whose social benefits exceed their costs should be made and that those investments for which this is not the case should not be made. Insofar as deciding what is efficient is concerned, the private or external nature of those benefits and costs is beside the point.

But these distinctions do affect actual behavior, and HCT takes great interest in how. We begin with the internal/external distinction. HCT argues that individuals and families have strong incentives to make investments in human capital that yield private benefits to them but no incentive to make investments in human capital that yield benefits to others. The immediate implication is that private actors will underinvest in forms of education that are not subsidized and yield substantial external benefits. After all, Junior will go (or be sent) to college if the private benefits to him and his family exceed the costs to him and his family. Why, it is reasoned, would they spend scarce resources to cover the costs of an education that primarily benefits other people? This understanding of the limits of privately elected and financed education is widely accepted by academic economists of all political persuasions. In a 1955 lecture, Milton Friedman allowed that "general education for citizenship" carries sufficiently important externalities (what he calls "neighborhood effects") that mandating minimal education levels and providing state funding for it makes sense (reprinted in Friedman 2002, 31).

Conversely, just as the absence of government subsidies can lead to the underprovision of varieties of education that carry big external benefits, the presence of subsidies can lead to the overprovision of education that provides mainly private benefits. This is because subsidies ensure that families pay less than the full social cost of education, which will lead some students to obtain privately beneficial education even if these benefits do not exceed the social cost of education. Thus, subsidizing education that carries primarily private pecuniary benefits raises the prospect of "over-education." Friedman notes this as well, arguing strongly against subsidizing "vocational" education that "increases the economic productivity of the student but does not train him for either citizenship or leadership" (2002, 88). It is therefore immediately obvious that, through the lens of this theory, what one considers the proper role of the state in education finance depends critically on whether one considers the benefits to be primarily private or external. If one believes, as we and most humanities scholars do, that most college courses can change people's minds in ways that yield external benefits, a serious role for the state is implied. On the other hand, if one believes that accounting courses simply prepare students to ply a more highly valued trade, then no subsidies to accounting schools are required. As noted, humanities scholars tend to be quite firm that universities should teach philosophy to accountants.

Next, with a focus on the private benefits of education, consider the pecuniary versus nonpecuniary distinction. This has large positive implications as well. If the education in question yields a stream of monetary rewards—typically higher wages—then it may be feasible to finance it privately, with cash-strapped students going into debt and using these future cash flows to pay off student loans. However, if the benefits are largely nonpecuniary, then private creditors may be less willing to finance student loans. In this case, replacing scholarships with private and debt financing will lead some less wealthy families to skip college altogether or to invest in cheaper forms of higher education whose benefit is more pecuniary. Moreover, academic majors whose benefits are disproportionately nonpecuniary are likely to see declining interest overall, and also from less wealthy and debt-dependent students, relative to majors offering disproportionately pecuniary benefits. As noted in section 10.1, and exemplified by the epigraphs we selected, the enrollment trends and the rhetoric around the humanities are quite consistent with a society obsessed with pecuniary benefits losing interest in the humanities—precisely the disciplines that seem to specialize in nonpecuniary benefits.[17]

Nonpecuniary benefits also pose a particular challenge to the use of efficiency as a normative criterion. It is nonpecuniary benefits whose monetary values vary with the beneficiary's wealth (recall: the rich philistine may place a higher monetary value on a poetry class than the poorer culture aficionado). When notions of fairness require equal access to these benefits, doing what is efficient will not generally be fair. Pecuniary benefits tend to suffer less from this type of subjectivity, because the market assigns them dollar values that are at least somewhat independent of the student's socioeconomic background. For example, a substantial share of what employers are willing to pay a worker surely depends on their contribution to production. And that contribution may have a lot less to do with how wealthy their parents are than does their willingness to pay for a poetry class. It is therefore easier to think of equality of opportunity as something that student loans and the market will deliver if the benefits of education are mostly pecuniary. When they are nonpecuniary, the markets will often replicate educational inequality across generations.

Putting these fundamental results of HCT together reveals in stark terms the limits of privately financed higher education. It is expected to reliably deliver efficient educational outcomes only if all the benefits of higher education are private and pecuniary, and if purveyors of student loans finance every educational investment for which these benefits exceed the costs. And those efficient outcomes are not likely to conform to any notion of fairness whenever education offers substantial nonpecuniary benefits. Thus, without public support, education that yields external or nonpecuniary benefits will be underprovided and education that yields disproportionately nonpecuniary benefits becomes an upper-class luxury. Because, as humanities scholars remind us, college yields many nonpecuniary benefits, totally private financing is expected to yield a separate and unequal higher education system.

This chapter's first two epigraphs suggest that these points have been entirely lost in public discourse. If one believes that the benefits of a good humanities education are disproportionately external, then humanities training should be publicly supported and subsidized relative to, say, STEM fields (science, technology, engineering, and mathematics). That humanities training is generally cheaper than STEM training adds to the argument for charging humanities majors lower fees. Instead, the advice from the nation's then top policymaker and his nearest rival appears to validate market outcomes that theory tells us are likely inefficient in the first place because college education does carry substantial external benefits.

The key oversimplifications of the version of HCT appearing in public discourse are by now clear: the external and nonpecuniary benefits are forgotten and efficiency is the only normative criterion applied. These simplifications have supported the transition from neoclassical HCT to neoliberal praxis.

10.4. HOW STUDENTS OF ECONOMICS LOST THE PLOT

Humanities scholars and economists agree that education policy must promote the delivery of education for its nonpecuniary and external benefits.

Section 10.3 established that, in order to apply HCT to this end while keeping both efficiency and fairness in mind, one simply must know how much of the benefits of an education are likely to be private/external and pecuniary/nonpecuniary. Education economists have every reason to attempt to figure this out. In this section we argue that they do so quite seldom, and that this is because of the formidable data-gathering and inferential problems involved. We also make the argument—one cannot prove such things—that a complete dependence on quantitative evidence has reduced economists' attention to those outcomes that are difficult to quantify.

We also argue that this state of affairs has led policymakers to make the simplifications we highlight, precisely because many of them have just enough economics training to form impressions of what economic research says about education finance but not enough training to recognize that these impressions are only half the story. However, before doing so, we make the case that many (probably most, and certainly the best) academic economists have always taken the external and nonpecuniary benefits of education seriously, even as they have lamented the difficulty of estimating them.

10.4.1 Economists Believe in All the Benefits of Education, but Struggle to Capture the Nonpecuniary and External

Early human capital theorists clearly did both of these things. Becker (1994) emphasizes that the principal purpose of his theory is to explain observed behavior with respect to schooling, which he emphasizes depends upon both "monetary" and "psychic" (i.e., private nonpecuniary) returns. He argues that the psychic gains, though difficult to estimate, are real and belong in any conceptual framework. He also emphasizes that normative work cannot lose sight of the external benefits of schooling, and indicates that, at an upper bound, these would be almost as large as the private ones. Nevertheless, he is forthright about his inability to measure either external or nonpecuniary benefits with certainty. These views were standard at the time.[18]

Broader questions of human capital investment took something of a backseat in the late 1970s and the 1980s as the West underwent major economic changes and stagflation. In conjunction with large postwar, publicly funded college expansions, weakening labor demand led to concerns over the declining wage returns of going to college—a trend brought to public attention by Richard Freeman in *The Overeducated American* (1976). This concern about the temporarily weak labor market experiences of young college graduates fed into a literature on "overeducation."

Overeducation is a potentially misleading term—it suggests that it would be efficient to reduce educational effort, even though overeducation studies measure only (some of) the private, pecuniary benefits of education. Some seminal papers in this literature are explicit about this discursive problem (Rumberger 1987), and those that are not are circumspect about not calling for reduced educational effort (Duncan and Hoffman 1981; Kiker et al. 1997).

The external benefits were back by the end of the decade. The "new growth theories" of the late 1980s and 1990s assumed large external benefits of human capital for the development of technological know-how (Arrow 1962) to develop models that revolutionized macroeconomic theorists' view of the role of education in economic growth (Aghion and Howitt 1998; Azariadis and Drazen 1990; Lucas 1988; Romer 1990). These authors claimed that their models could explain features of the historical growth record that were previously unexplainable.

Attention in the late 1990s shifted to testing these new growth theories. One might think of this as a search for direct evidence of the dark matter postulated by new growth theory—external benefits might help explain the growth record, but could their existence be confirmed?

One particularly irksome finding from this literature was that comparisons across countries over time did not yield clear evidence of production externalities due to human capital accumulation. For if these externalities were large, then why did the explosion in education in developing countries in the 1980s and 1990s not lead their gross domestic products to explode as well (Pritchett 2001, 2006)? Two possible explanations hold that the lack of a relationship between educational and economic expansion is due to problems of quantitative measurement. The first is that analysts cannot accurately measure changes in national education levels, so that the relationship between educational expansion and growth is obscured by measurement error (Krueger and Lindahl 2001). The second is a general problem with measuring the effects of education, or just about anything else, on economic growth: the number of possible explanations for national growth experiences is large relative to the number of countries in existence, so that teasing out the relevance of each possible explanation is statistically challenging (Sala-i-Martin 1997). Either explanation implies that education can carry large external benefits under the right conditions but that these benefits are difficult to render observable in the quantitative record.

The profession underwent an empirical turn in the early twenty-first century. This was motivated by several factors, two of which are particularly relevant to economists' understanding of education: big improvements in econometric practice cast doubt on old ways of parsing quantitative evidence of causality (Angrist and Pischke 2008, 2010; Imbens 2010) and economists increasingly borrowed and tested ideas from other disciplines (particularly psychology, sociology and political science) that probed the limits of human rationality (Akerlof and Shiller 2015; Leonard 2008). This newfound intellectual openness led to renewed interest in the nonpecuniary benefits of education. For if we are not totally rational, how we think becomes malleable and changing it becomes interesting. Economists have found lots of evidence to suggest that education does change how people think and what they value.[19]

Some of these changes can be negative, hindering peace and democracy, and so these researchers discuss the nonpecuniary "effects" of education, only occasionally describing these effects as "benefits." However, one might further divide these nonpecuniary effects into sociopolitical effects, some of which are negative, and personal effects, which tend to be overwhelmingly positive and so constitute nonpecuniary benefits.

For example, Oreopoulos and Salvanes (2011) show that education reduces people's tendencies to smoke and to support the use of corporal punishment on children. They go on to demonstrate many more personal nonpecuniary benefits: greater happiness, more stable families, job satisfaction, occupational prestige, health, and ability to delay gratification. They also point out that most students actually enjoy college. Although these effects are less transformative than those envisioned in the humanities tradition, and more contingent on other political institutions, their size confirms humanities scholars' claims that nonpecuniary and external effects are important and are intimately tied to the educational process.

Three features of this economic literature on nonpecuniary benefits are noteworthy in a book on the effects of quantification. First, it tends to be concerned with primary and secondary rather than tertiary education. This is unsurprising, given that the sociopolitical effects of education depend critically on not just what is taught but how it is taught (Algan et al. 2013). The much greater diversity of educational experiences at the college level makes the nonpecuniary benefits of college education contingent, variable across students, and therefore more difficult to measure.

Second, with one exception (McMahon 2009), we know of no scholars who attempt to put monetary values to the nonpecuniary benefits of higher education. This is noteworthy because, from the perspective of HCT, doing so is necessary in order to answer the most fundamental questions about what arrangements for financing college are likely to be efficient.[20] Indeed, Oreopoulos and Salvanes (2011) are the only authors widely cited by economists who compare the pecuniary and nonpecuniary benefits of education, and they do not do so by putting monetary values to nonpecuniary benefits. Instead, they treat happiness—a nonpecuniary benefit—as the true end, and income gain merely as a means toward it. They find that *only one-quarter* of the effects of education on happiness can be attributed to the higher earnings of educated workers.

Third, whenever economists are able to measure the sociopolitical effects of education the papers detailing their findings are accepted by editors at the most widely read, general-interest journals. This confirms that the profession as a whole is deeply interested in what education does, beyond its pecuniary private benefits. The problem, as we have hinted and shall now demonstrate, lies in the methodological difficulty of capturing the nonpecuniary and external benefits, as a result of which they are not often discussed.

10.4.2 Distortions in Column Inches Lead to Communication Breakdown

So why do so many consumers of economic analysis not share economists' appreciation for education's nonpecuniary and external benefits? The proximate cause, we suggest, is that undergraduate and graduate students of economics, as well as those studying some economics in the fields that lead to jobs making and influencing policy (public policy, law, business, journalism), tend to be overwhelmingly exposed to materials that focus on the pecuniary private benefits.

One set of materials of particular importance is introductory economics textbooks, which often state explicitly that they aim to teach students to think like economists. For most students of economics who do not major in the subject, such books provide a critical first and last impression of what benefits of college economists care about. We therefore analyzed the treatment of the benefits of education in the five introductory general economics textbooks on our shelves. They leave the distinct impression that the benefits of education are mostly private and pecuniary.[21]

Econometrics textbooks and the exercises they contain provide most economics majors their first exposure to how the effect of education can be measured. For graduate students, they are a central part of the canon, illustrating the technical and inferential problems one must learn to solve to advance in the profession. Again, we pulled every general econometrics textbook off our shelves (thirteen books in total) and searched each one from cover to cover for examples or practice problems related to the economics of education. Not one of them included any material on estimating its nonpecuniary or external effects.[22]

Finally, with fundamentals secure, economics students hone their craft in the seminar room, watching graduate students and faculty present papers. Many of the papers on the economics of education to which students are exposed in seminars (and certainly the vast majority of those that students write) are not path-breaking, but rather offer modest improvements in measurement and interpretation. Such papers are usually not deemed suitable for general-interest journals. Many are published in the *Economics of Education Review* (*EER*), the leading journal dedicated to the subject.

Treating the *EER* as indicative of what the average education economist spends their time researching, we examined sixty-two systematically sampled articles appearing in the *EER* from its founding in 1982 until 2017, and examined how much attention they pay to the nonpecuniary and external benefits of education. The results, presented in box 10.1, confirm that education economists routinely recognize the importance of these benefits but do not actually study them very often.

BOX 10.1. Content analysis of papers appearing in the *Economics of Education Review*

METHODOLOGY

We analyzed the 7–12 articles (excluding book reviews) pertaining to higher education that appeared in the journal in each of eight quinquennial years (1982, 1987, 1992, . . . , 2017), or 62 articles in all. We trained two research assistants to read and code the articles according to how seriously the authors considered each of 20 specific benefits of college. Benefits included, for example, higher wages (a private pecuniary benefit), greater happiness (a private nonpecuniary benefit), and increased regional wages (an external pecuniary benefit). Each of these 20 benefits was mapped to a cell in a 3×3 grid (private/external/ambiguous and pecuniary/nonpecuniary/ambiguous). Coders were also instructed to consider any benefits beyond these 20 that might belong in any cell. Each paper was assigned scores reflecting how seriously they took each of these benefits, and the maximum among the benefits corresponding to a cell became the score for the paper in that cell. Intercoder reliability was high, and differences in coding choices were resolved between the research assistants at the start, with input from one of the authors.

We also coded articles according to whether they treated the distribution of education or its benefits among people of different social groups (race/ethnicity, gender, income) as a normatively important outcome, and if so, whether they did so explicitly or implicitly.

We coded the seriousness with which each benefit was treated as follows: 0—the benefit is not mentioned; 1—it is mentioned but dismissed as unimportant; 2—it is treated as potentially important but is not a focus of the study; 3—it is a central focus of the study, as evidenced by the fact that it is explicitly theorized, measured, or studied in some other way.

FINDINGS

Only 40 of the 62 higher-education papers we coded treated any economic benefit of education as potentially important (score = 2 or 3), and only 25 actually studied such a benefit. The 22 papers that do not explicitly treat any benefits as important focus on such matters as the cost of higher education and the makeup of the student or faculty body.

Of the 40 papers taking one or more benefit of college seriously, 36 treat a private pecuniary benefit this way, 11 do so for an external pecuniary benefit, and 9 do so for a private nonpecuniary benefit. Only one paper explicitly treats an external nonpecuniary benefit as important. A total of 21 out of 40 papers treat at least one benefit that is not private and pecuniary as important. Thus, so long as one does not require an analyst to actually study a particular benefit of higher education, many papers in the *EER* acknowledge the importance of nonpecuniary and external benefits. Nevertheless, it is also true that more *EER* papers treat private pecuniary benefits as important (36) than treat benefits that are not private pecuniary as important (21).

However, if we consider only the 25 papers that actually study at least one particular benefit (score = 3), 22 of them study a private-pecuniary benefit, only 7 study a benefit that is not private-pecuniary, and only 3 study an external benefit. Thus external and nonpecuniary benefits are mentioned fairly often but are seldom studied.

Our analysis of *EER* articles also confirms that many economists treat the distribution of the benefits of higher education within the population as an important normative outcome. Out of our 62 papers, 18 explicitly treat it as normatively important and another 4 do so implicitly.

Our analysis of *EER* articles also confirms that many economists treat the distribution of the benefits of higher education within the population as an important normative outcome. Out of our sixty-two papers, eighteen explicitly treat it as normatively important, and another four do so implicitly.

Thus, our analysis of *EER* content confirms that education economists are generally aware of a wide variety of benefits of college and care about who these benefits accrue to. Yet they produce papers skewed toward the private pecuniary benefits that privately funded college tends to produce, and away from the external and nonpecuniary benefits that humanities scholars emphasize. A more casual empiricism suggests that similar results would be replicated in a study of *Education Economics,* the other journal dedicated to the economics of education.

10.4.3 Disparities in Column Inches Are Driven
in Large Part by Evidentiary Standards

What explains the limited attention paid to the nonpecuniary and external benefits of education in textbooks and specialized journals? The reasons appear to be different for external and nonpecuniary benefits, but in both cases they come down to an insistence on specific forms of quantitative evidence.

In the case of external benefits, the problem is one of degrees of freedom. To illustrate, consider "productivity spillovers"—the idea that workers are more productive, and are therefore paid more, when they live and work around more educated people. Spillovers operate and can become apparent in variation only across larger units of analysis (e.g., regions, countries). This means working with a smaller effective sample size—most countries have millions of workers, but at best, dozens of regions. Moreover, with this smaller effective sample size, one must also show that correlation between an individual's wages and the education of their peers is not explained by some other factor that varies across regions (e.g., school quality, legal institutions, structural features of the economy, geography, historically determined locations of industries, etc.). The number of regions is often small relative to the number of such explanations, reducing analysts' power to rule particular explanations in or out.

A further issue is that the education levels of regions are not randomly determined. Workers respond to accelerating economic development and change by increasing their investments in education. Thus correlations between regional education levels and individual workers' productivity levels are difficult to interpret causally. Prestigious economics journals today do not accept papers that are unable to credibly identify causation or offer Boolean, defensible conclusions. Textbooks, dedicated as they are to presenting what is well established in the discipline, similarly do not report on findings that are open to debate.

We base our argument that many of the external benefits tend to be ignored due to degrees of freedom problems on several pieces of evidence. First, section 10.4.1 reviewed several papers accepted by prestigious journals which do document the *potential* for external nonpecuniary effects of education (higher tendencies to trust, vote, volunteer, etc.), but each of these potentials is observable in the behavior or attitudes of a single person. They do not directly demonstrate that these tendencies actually result in more educated regions or societies doing better. The literature actually demonstrating that is small.[23] Strong circumstantial evidence at a disaggregated level and weak direct evidence at the aggregate level are precisely what one would expect when degrees of freedom bind on aggregate inferences.

Second, we have direct evidence. Pritchett's (2006) review of the literature on the production externalities of education explicitly raises three difficulties which add up to serious power problems: What exactly is meant by a "regional education level," and how it should be measured? What other variables should we correct for (or not) as we compare across regions (and how should these be measured)? And are there enough comparable regions to estimate relationships precisely? Others note that inferring the direction of causality is also hard—high local education levels may be the consequence, not the cause, of productive cities (Machin and Vignoles 2018).[24]

Quantification, and particularly an insistence on evidence from large-N analysis, is implicated here. Political scientists, whose job is more often to explain collective rather than individual outcomes, face the degrees of freedom problem more regularly than economists. They have historically dealt with it creatively, turning to well-chosen case studies and process tracing in order to tease out causal connections where statistical analysis hits its limits (Collier 2011; King et al. 1994). These approaches very rarely appear in economics journals, and case study evidence is often dismissed in the profession as merely suggestive. Our point here is not that large-N analysis is unhelpful. Rather it is that the limited large-N evidence of ubiquitous external productivity benefits is not evidence that these externalities do not exist or are unimportant everywhere, and that when the consequences of false negatives are serious, other forms of evidence—often qualitative—must be taken seriously.

With nonpecuniary benefits, the problem may simply be that the data is rare. Data on wages is abundant, data on civic engagement is not. This explains why the *EER* does not feature many articles on nonpecuniary benefits—having something credible and new to say about them is gold and moves papers to general interest journals.[25] Data constraints may also explain why econometrics textbooks do not reference nonpecuniary benefits: much as the *New York Times Cookbook* does not contain recipes for abalone given its limited availability in supermarkets, illustrations of the pitfalls of working with data on civic engagement are unlikely to be of much value to users of econometrics. Exactly why introductory economics textbooks do not discuss nonpecuniary benefits is less clear. After all, there is no controversy in the profession nowadays over their existence, and the number of datasets on happiness has grown enormously. The best quantitative evidence of their importance is recent. If our conjecture is correct, then the next generation of textbooks should resolve these issues.

If they don't, this would be an indication that the blind spots we accuse policy makers of extend upstream into professional economics itself.

10.5. CAVEATS

We have argued that humanities scholars' and economists' approaches to higher education are mostly compatible in theory but that the difficulty of providing quantitative evidence on many of the benefits of education has narrowed and distorted understandings of the role of higher education in the policy community and among the wider public. Several limitations to this argument are worth noting.

First, quantification is not the only reason that policymakers have forgotten about nonpecuniary and external benefits. Another reason is ideological: as the movement sometimes called neoliberalism largely discredited government management of services in US and UK political cultures, treating higher education as a market good became a general common sense. Accepting that there are external benefits requires public spending, and raising tax revenue in an antigovernment era is difficult. Increasing racial diversity in universities may have dampened the enthusiasm of the white majority, and of those elites and working-class taxpayers who are least likely to attend public universities, for funding them through general taxation. Professional incentives in policy circles therefore ran against putting nonmarket functions of higher education into debate. Furthermore, as wealth and income inequality increased, attention turned to financing arrangements that could mitigate the impact of declining public funding per student on access to universities among lower-income students, and the viability of these arrangements hinged on calculations of pecuniary return on investment. In short, ideological, political, and social changes also encouraged the neglect of nonpecuniary benefits. As is widely discussed, economics has played a role in these changes (Brown 2015).

Second, HCT is not the only influential theory of education developed by economists. Signaling and screening theories show that college can be rewarded if graduation reveals workers' otherwise hidden capabilities, even if the education adds nothing to those capabilities (Arrow 1973; Spence 1973). Thus, college's primary role could be to sort workers into better and worse employment and social opportunities. On this theory, if cheaper sorting mechanisms are available, then less higher education is required. Caplan (2018) for example, argues that education's main function is to signal status. Worse, some argue, public financing of universities (rather than of students, through financial aid) would tend to increase social stratification by permitting students from more advantaged families to monopolize these opportunities (Reeves 2018).

These ideas have indeed weakened the social case for higher education and for funding it publicly. Yet, here, too, there is no reason to believe these views reflect the philosophical predisposition of economics as a discipline. The most stringently peer-reviewed empirical studies do find evidence consistent with education carrying signaling benefits but do not leap to the conclusion that it carries no other benefits (e.g., Bedard 2001; Lange and Topel 2006). Caplan's assessment that only 20 percent of the wage gains from education are due to actual improvements in how students think does not reflect a consensus in the quantitative empirical literature.

Third, our argument that research economists' focus on pecuniary private benefits reflects the relative availability of quantitative evidence, rather than an intellectual bias, might be too optimistic. Perhaps our reference sample of economists is more broad-minded than typical. Settling that issue would require a representative survey of the profession. To our knowledge, no such study exists.[26]

10.6. CONCLUSIONS

Our analysis allows us to compare economic and humanistic understandings of the effects of college. To generalize, humanities scholars value higher education because it creates nonpecuniary benefits. These center on large improvements in how well people think and what they can think competently about. In general, humanities disciplines treat pecuniary effects as spillovers of higher learning rather than its direct aims. Thinking with greater rigor and scope will make it more likely that the graduate will find better-paying work, but that is not higher education's direct purpose.

In contrast, economists have focused on the pecuniary effects of university education. The key reason for this, we've argued, is the profession's overwhelming preference for quantitative over qualitative evidence. Many of the nonpecuniary effects of education are not directly observable, and even if they are—by sociologists, historians, cultural critics, and others— they are often not measurable. Quantitative assessments of the benefits of education are therefore likely to be incomplete, and so to discount what the humanities have considered important.

We thus have two traditions that have analyzed the effects of higher education while standing back to back. Economists focus almost exclusively on one of the four types of educational benefits—the private pecuniary— while acknowledging the existence and importance of the others and measuring them when the opportunity arises. Humanities scholars have focused almost exclusively on two types of nonpecuniary benefits, the private and the external (what they might call individual and social). When these groups talk about the benefits of higher education, they are not so much canceling each other out as focusing on different benefits.

The authors have found that when we do talk about the same things, we tend to agree on several points. First, it is simply wrong to base an understanding of the benefits of higher education on analysis of the private pecuniary benefits alone. This can be asserted quite generally: no accepted economic (much less humanities) analysis of what education does in the world supports the view that the nonpecuniary and external benefits are zero. Second, and as a consequence, requiring the recipients of higher education to shoulder too much of the cost of their schooling will result in socially damaging underinvestments in higher education. We do not agree on how much is "too much," but do agree on a third point: that the systematic shifting of the costs of higher education onto students has gone too far in countries like the United States and the United Kingdom that have embraced market liberalism and should be reversed. This will no doubt be contested as insufficiently grounded empirically, and for the reasons we enumerated in section 10.4.3, we are not sure the debate is empirically resolvable.

Rather, our shared willingness to risk erring on the side of "overinvestment" in schooling rests on a further point of theoretical agreement. If we require students to pay for college out of pocket, students with less money will be less able to afford it. Students will also be less able to pay for—and therefore less able to acquire—education that delivers nonpecuniary benefits. This concern is underscored by the evidence cited at the opening of this chapter, of stubborn racial stratification across universities and declining humanities enrollments as the privatized funding model comes into its own.

Finally, we found agreement on two required conceptual improvements to the debates over funding models. One is that instead of casting discussions about college affordability in terms of tuition, or of all direct costs (tuition, books, and fees), these discussions should consider the full opportunity costs of attending college. This would include the students' additional costs of food and shelter and foregone earnings. Human capital theory reminds us, and the interdisciplinary university studies literature confirms (Goldrick-Rab 2016), that all of these costs affect families' and students' decisions. Consideration of these costs and their consequences for cash flow should increase support for more generous and progressive financial aid policies. This improvement is wholly compatible with mainstream economics.

The other required improvement is to acknowledge, properly conceptualize, and attempt to account for (in comparable ways) the nonpecuniary and external benefits of different types of education. This is a very large job, where contemporary humanities scholars could do more than they have, but economists also need to contribute to the theory of nonpecuniary effects.

Were these improvements to be made, there would be two clear benefits. Fewer students would be excluded from studying what they wish to because of their lower ability to pay. And students whose majors benefit society more in nonmonetary than in monetary ways should have to pay less—and enjoy greater public esteem than is the case today.

NOTES

1. Definitions of HCT vary across authors. For this chapter, it refers to the range of ideas covered in Becker's (1993) original microeconomic treatise on the subject. In *The Death of Human Capital*, Brown et al. (2020) use the term much more expansively to critique a suite of ideas that we would refer to as human capital fundamentalism. One of these ideas is that education's primary purpose is to increase labor productivity and earnings; another is that increased availability of education creates its own demand. We agree with their assessment that both ideas are ubiquitous and wrong. However, we treat the first idea itself as a simplification of what HCT actually says, and the latter as a wishful macroeconomic extrapolation from the microeconomic framework we refer to as HCT. These are merely differences in definition, not substantive intellectual disagreements.

2. As three scholars of the subject observed,

> Although the higher university faculties—the traditional term for law, medicine, and theology—struggled to maintain enrollments, none struggled more than the lower faculty, the traditional term for the philosophy or arts faculty. Philosophy faculties prepared students for study in one of the higher faculties. They were a propaedeutic for students and professors, who sought the higher pay and prestige of a position in one of the higher faculties. Over the course of the eighteenth century, university students began to forgo the philosophy faculty entirely and enroll directly into one of the higher faculties, widely considered more practical and professionally oriented. (Menand et al. 2016, 84)

3. We are used to American statements about the university's double function, as in Johns Hopkins University founder Daniel Coit Gilman's assertion that universities answer to "the double test, what is done for personal instruction, and what is done for the promotion of knowledge." There was nothing double about this function to the German theorists. See Daniel Coit Gilman, "The Utility of the University" (1885), in Menand et al. (2016). Humboldt was particularly clear that knowledge could and should be created in the act of teaching itself. If a lecture was not also pushing back against limited formulations and troubling obscurities of thought and expression then it was not really teaching at all.

4. James A. Good (2014, 2) has a clear summary of Bildung that stresses its roots in Herder's "neohumanism," which rejected Lockean epistemology and standard English-language philosophical distinctions:

> In a series of works written over a period of almost fifty years, Herder developed and defended the conception of philosophy that is at the very heart of the German *Bildung* tradition. The titles of some of these works are revealing: *How Philosophy Can Become More Universal and Useful for the Benefit of the People* (1765), *This Too a Philosophy of History for the Formation of Humanity* (1774), *Ideas for the Philosophy of History of Humanity* (1784–91), and *Letters for the Advancement of Humanity* (1793–97). As these titles suggest, Herder believed philosophy must have a practical result, which can be summarized as human growth, and that philosophical ideas have to be understood within their social and historical context. Similar to the Renaissance Humanists, Herder believed that the proper study of man is man, and thus sought to displace academic philosophy with philosophical anthropology. For Herder, philosophy is, quite simply, the theory of *Bildung*; more precisely, philosophy is the theory of how the individual develops into the sort of organic unity that will constantly work toward the full development of its talents and abilities and that will drive social progress or social *Bildung*. For Herder, properly understood, philosophy must transform individuals and, at the very same time, it must have a broad social impact. John Zammito rightly asserts that the conception of philosophy Herder defended carried "forward from Herder to Wilhelm von Humboldt and G.W.F. Hegel, to Friedrich Schleiermacher . . . to the Left Hegelians . . . and Wilhelm Dilthey: the tradition of hermeneutics and historicism."

5. The quotation is one of André Malraux's characters' answer to the question, "How can one make the best of one's life?," cited in Hofstadter (1962, 28). Hofstadter's distinction between intelligence and intellect reflects the ongoing influence of Prussian university theory's contrast between the outputs of the higher and lower faculties, respectively, and between applied and self-determined thought. Hofstadter defined "intellect" as "disinterested intelligence, generalizing power, free speculation, fresh observation, creative novelty, radical criticism" (27). He also distinguished it from professional activity as such, which is a "stock of mental skills that are for sale."
6. Here is Fichte again:

> The point of studying is not, after all, to enable someone to spend his whole life spouting off what he had learned for an exam years before. The point is for him to be able to apply what he has learned to the situations and predicaments that come up in life, and thus transform what he has learned into action: not merely repeat it but make something else from it and with it. In other words, here as elsewhere, the ultimate goal is not knowledge but the art of using knowledge. (Menand et al. 2016, 72)

7. Thanks to Jan-Melissa Schramm for presenting these passages in a talk at the University of Cambridge, June 20, 2018.
8. Angell's peroration sounded like this: "In justice then, to the true spirit of learning, to the best interests of society, to the historic life of this State, let us now hold wide open the gates of this University to all our sons and daughters, rich or poor, whom God by gifts of intellect and by kindly providences has called to seek for a liberal education" (1879, 19).

Angell refers positively to the previous year's commencement address, George Van Ness Lothrop, "A Plea for Education as a Public Duty" (1878), which offers the following summary list of main points:

1. That education is the duty of the free State, because necessary to its safety. 2. That the higher education is but a necessary part of any sufficient system of education. 3. That, historically, Michigan is pledged to provide for and maintain this complete system of education. 4. That it enhances the power and the welfare of the State in peace and in war. 5. That only by this means can we hope for the conquest over the forces of nature so necessary to the safety of society. 6. This University is not only the consummation of what has been done, but the promise of what shall be done by Michigan in the great duty of offering her children the privileges of the most liberal instruction. (Angell 1879, 19)

9. See, for example, Henry's (2017) account of the Yale 1967 conference on "Black Studies in the University" (35–37).

10. It must be noted that Coates (2015) describes his practice of study at Howard University as the continuation of his study in his parents' home, his friends' homes, in bookshops and record stores and magazine stands:

Contrary to [Bellow's] theory, I had Malcolm. I had my mother and father. I had my readings of every issue of The Source and Vibe. I read them not merely because I loved black music—I did—but because of the writing itself. Writers Greg Tate, Chairman Mao, dream hampton—barely older than me—were out there creating a new language, one that I intuitively understood, to analyze our art, our world. (44)

Coates's descriptions of Howard continue from that point:

When I came to Howard, Chancellor Williams's Destruction of Black Civilization was my Bible. Williams himself had taught at Howard. I read him when I was sixteen, and his work offered a grand theory of multi-millennial European plunder. The theory relieved me of certain troubling questions—this is the point of nationalism—and it gave me my Tolstoy. I read about Queen Nzinga, who ruled in Central Africa in the sixteenth century, resisting the Portuguese. I read about her negotiating with the Dutch. When the Dutch ambassador tried to humiliate her by refusing her a seat, Nzinga had shown her power by ordering one of her advisers to all fours to make a human chair of her body. That was the kind of power I sought, and the story of our own royalty became for me a weapon. My working theory then held all black people as kings in exile, a nation of original men severed from our original names and our majestic Nubian culture. Surely this was the message I took from gazing out on the Yard. Had any people, anywhere, ever been as sprawling and beautiful as us? I needed more books. (45–46)

On *study*, see Harney and Moten (2016).

11. The attack on the humanities in particular and the university in general took the form of the "culture wars." See part I of Newfield (2008).

12. Leading exponents of antifoundationalism include Stanley Fish in literary criticism (1982), Richard Rorty (1989), Butler (1990), Donna Haraway (1990) in a different idiom, and Sandra Harding (2004), and many others.

13. Wood (2015) offers a very useful summary of de Man's deconstruction:

> De Man's enemy is our habit of submitting "uncritically to the authority of refer-
> ence", of assuming that language always points away from itself and points clearly.
> We are devoted to "the myth of semantic correspondence between sign and referent",
> caught up in "the fallacy of reference". These are all grand overstatements, as de Man
> himself indicates when he says, "the notion of a language entirely freed of referential
> constraints is properly inconceivable", but they have a job to do, and they allow us to
> see why de Man thinks that "far from being a repression of the political, as Althusser
> would have it, literature is condemned to being the truly political mode of discourse".
> Literature in this view is what helps us to think twice, a confirmation of Brecht's view
> that all thought is on the side of the oppressed. Or at least is against the myths and fal-
> lacies the authorities want us to believe are true. "There should be no fundamentalists
> in criticism", de Man says.

14. As a crude indication, a Google Scholar search yields roughly 50,000 articles
featuring the phrase "human capital theory." The number of articles drawing on the
theory itself is probably an order of magnitude greater.

15. See note 1 above for a discussion of broader uses of the term than ours. In
narrower usage, the term designates theories that hold that educated workers are
paid more because education directly increases their productivity (e.g., McGuinness
2006).

16. One may counter that such distinctions are not made when conducting a
cost-benefit analysis for purposes of determining public spending, and this is often
true. But when the theoretical argument is made that using market mechanisms can
coordinate educational investments efficiently, it is beyond doubt what is implied,
for the maid's daughter will, in the market, express a willingness to pay for schooling
that is constrained by her family's finances and access to credit.

17. These class-differentiated defects of rising costs on the demand for different
majors may not translate into observable differences in who graduates from which
majors. As departments promising more pecuniary benefits are more likely to restrict
entry into their majors, middle-class students are often more successful than their
working-class counterparts in their efforts to exit the humanities.

18. For example, Witmer's (1970) summary of the state of the field cites several
authors he deemed highly influential at the time whose work draws attention to the
external and nonpecuniary benefits of college. Like Becker, he notes that these returns have been difficult to estimate, and indeed, that whether they can be estimated
is a matter of some controversy. Nevertheless, he concludes that "the following external benefits flow from higher education and ought to be considered by any prudent
decision maker: higher levels of political participation, higher levels of tax revenue,
higher levels of mobility, intergenerational gains through the informal education
the children of college graduates receive at home, lower levels of unemployment,
and leadership in maintaining political democracy and a free market system" (519).
Shortly after, in a short book intended to convey ideas from human capital theory
to British educators, Woodhall and Ward (1972) provide a detailed analysis of the
private pecuniary returns to education, noting, "Even if it were possible to measure
all the economic benefits of education satisfactorily . . . and this is far from the being

the case . . . there would still remain the question of how to take account of non-economic benefits, and how much priority to attach to the consumption, as opposed to investment aspects of education" (35).

19. For example, Lott (1999) argued that totalitarian governments value indoctrination over real education, and so are more apt to invest in public education than are democratic ones. This led several economists to test the assumed prior—that education can really influence political thought. These authors found that education could create a sense of national identity (Clots-Figueras and Masella 2013), that it can lead to reduced willingness among girls to accept domestic violence and political authority, but greater willingness to consider political violence legitimate (Friedman et al. 2016), and an acceptance of the party line on governance, political institutions and economic institutions (Cantoni et al. 2017). More positively, Dee (2004) shows that US students drawn into higher education by the proximity of community colleges are more likely to vote and to register to vote, while Lochner (2011) shows that educated citizens are less likely to engage in criminal activity. Glaeser and colleagues (2007) argue that education is complementary to democracy. They review other literature showing a significant effect of schooling on civic participation, and also show that initially more educated countries tend to move towards democracy as time passes. They point out that more educated Americans are not just more active politically, but are more socially engaged—more likely to attend church, classes or seminars, more likely to work on community projects, and less likely to flip off other drivers. This leads them to suggest that in democratic societies, education socializes rather than indoctrinates. This complements work showing that when (and only when) education involves group work and student participation, it tends to promote trust and the building of social capital (Algan et al. 2013).

20. It is, of course difficult to say why this happens. We have found no smoking guns. Perhaps this particular gap in the literature reflects the practical difficulty of deciding what any of these effects are worth in monetary terms. Or it may reflect a discomfort with producing monetary values of the nonpecuniary benefits of education that vary with the socio-economic status of the student involved. It could also reflect a prior for some economists that the benefits are not worth quantifying.

21. The only benefit of education that Samueleson and Nordhaus's classic (2005) textbook refers to is its effects on individual productivity. They come close to recognizing an external benefit in one oblique reference to government spending having paid for the education of some people behind successful inventions. Krugman and Wells (2006) also cover no benefit of education other than its contribution to economic growth and to individual earnings. Baumol and Blinder (2007) explain that people invest in education to obtain nonwage (private) benefits such as prestige, but nevertheless focus only on education's effects on earnings. Even Colander (2008), who emphasizes the ecumenicity, historicity, and institutionalist emphasis of his textbook, treats education purely in terms of its effects on production and earnings.

22. Ten books (Angrist and Pischke 2008; Cameron and Trivedi 2005; Davidson and MacKinnon 1993; Greene 2008; Gujarati 1992; Hsiao 2014; Kennedy 2008; Maddala 1977, 1992; Wooldridge 2002) included examples and exercises that involve estimating the private wage benefits of education. Of the remaining three, only one (Goldberger 1991) neglected to discuss this topic—and the other two (Johnston 1972; Judge et al. 1980) included no examples at all.

23. It is true that a modest literature does attempt to measure spillovers at the regional level (Glaeser and Saiz 2003; Moretti 2004; Shapiro 2006; Valero and Van Reenen 2019), but these exceptions prove our point.

24. Two quotes illustrate: "The empirical evidence to date produces little support for output externalities. This is . . . also in part because of low power, as confidence intervals of estimates of returns from aggregate data often include zero output impact, large negative estimates, [estimates equal to] the Mincer returns (zero externalities), and estimates consistent with large externalities" (Machin and Vignoles 2018, 638). "Imposing sufficient structure to interpret the results (of regressions using aggregate data) as a 'rate of return' to schooling raises a host of essentially insuperable problems and requires an active suspension of disbelief" (653).

25. One illustration of the problem comes from Dickson and Harmon (2011). This introduction to an *EER* special issue states that the papers it contains came from a conference targeting three underresearched areas in the economics of education, including the nonpecuniary benefits of education. They note "perhaps economics as a profession has allowed a major body of research on the non-pecuniary returns . . . to become dominated by the other social sciences" (1119). Yet, none of the papers in the issue actually addresses nonpecuniary benefits.

26. We are grateful to Heather Steffen for numerous excellent discussions, comments on drafts, and administrative support; to Claire Hunt and Swaroopa Lahiri for painstaking research assistance; and to Gabriele Badano, Zachary Bleemer, Elizabeth Chatterjee, Trenholme Junghans, Mukul Kumar, Greg Lusk, Laura Mandell, and Chris Muellerleile for excellent suggestions. This work was supported by the Independent Social Research Foundation (Cambridge group), the Mellon Foundation (Chicago group), and the National Endowment for the Humanities (Santa Barbara group).

Acknowledgments

This volume is an outcome of an unlikely collaboration between people who, while unified by a common theme, are otherwise separated by disciplines, institutions, and geography. So our greatest thanks go to those who brought us together in 2015. They are Simon Goldhill of the University of Cambridge, Louise Braddock of the Independent Social Research Foundation (ISRF), and James Chandler of the University of Chicago. They sketched the initial intellectual vision and negotiated the funding and institutional support at those two universities. Equally important were the relationships we built and the admiration we developed for each other in our many intense conversations in Cambridge and California. At the University of California, Santa Barbara (UCSB), John Majewski, Dean of the Humanities and Fine Arts, supported the higher education branch of the collaboration until that group could obtain extramural funding. The National Endowment for the Humanities (NEH) Collaborative Research Program (grant RZ-255780-17) offered primary research support for two years; group research is usually done on a purely voluntary basis in the humanities and we are grateful to the NEH for funding one of the rare exceptions to the rule. Also crucial was another rare exception: the University of California Humanities Research Institute (HRI) provided a residential fellowship to an overlapping group of scholars, which allowed eleven weeks of uninterrupted debate and reflection across disciplines; we are grateful to director David Theo Goldberg and to the HRI staff, particularly Suedine Nakano and Arielle Reed, for their organizational support. We are also deeply grateful for the patient administrative work of Michelle Maciejewska of the Centre for Research in Arts, Social Sciences and Humanities (CRASSH) at Cambridge, Stuart Wilson of ISRF, and Tracey Goss and Marcelina Ortiz of the Chicano Studies Institute at UCSB. Many brilliant scholars of quantification participated in our workshops and conferences and otherwise provided intellectual inspiration. We are especially indebted to workshop participants and research colleagues Ted Porter,

Mike Kelly, Havi Carel, Leah McClimans, Isabelle Bruno, Emmanuel Didier, and the late great Sally Engle Merry. Alan Thomas of the University of Chicago Press provided our volume proposal with a generous and insightful review. Ulrika Carlsson reviewed the entire volume for coherence and style with a remarkable rigor and speed, and made a major contribution to the volume's consistency and clarity.

References

Adams, Vincanne, Michelle Murphy, and Adele E. Clark. 2009. "Anticipation: Technoscience, Life, Affect, Temporality." *Subjectivity* 28, no. 1: 246–65.

Adapt Smart. n.d. Accessed February 26, 2021. http://adaptsmart.eu/home/.

Aghion, Philippe, and Peter Howitt. 1998. *Endogenous Growth Theory.* Cambridge, MA: MIT Press.

Ahlstrom-Vij, Kristoffer. 2013. *Epistemic Paternalism: A Defence.* London: Palgrave Macmillan.

Aizer, Anna, and Joseph J. Doyle Jr. 2013. "Juvenile Incarceration, Human Capital and Future Crime: Evidence from Randomly-Assigned Judges." National Bureau of Economic Research Working Paper 19102. https://www.nber.org/papers/w19102.pdf.

Akerlof, George A., and Robert J. Shiller. 2015. *Phishing for Phools: The Economics of Manipulation and Deception.* Princeton, NJ: Princeton University Press.

Alexandrova, Anna. 2017. *A Philosophy for the Science of Well-Being.* New York: Oxford University Press.

Algan, Yann, Pierre Cahuc, and Andrei Shleifer. 2013. "Teaching Practices and Social Capital." *American Economic Journal: Applied Economics* 5, no. 3: 189–210.

Allen, Danielle. 2016. *Education and Equality.* Chicago: University of Chicago Press.

Allen, Jonathan, and Amie Parnes. 2017. *Shattered: Inside Hillary Clinton's Doomed Campaign.* New York: Crown.

Allen, Myles. 2011. "In Defense of the Traditional Null Hypothesis: Remarks on the Trenberth and Curry WIREs Opinion Articles." *Wiley Interdisciplinary Reviews: Climate Change* 2, no. 6: 931–34.

Allen, Myles, Pardeep Pall, Dáithí Stone, Peter Stott, David Frame, Seung-Ki Min, Toru Nozawa, and Seiji Yukimoto. 2007. "Scientific Challenges in the Attribution of Harm to Human Influence on Climate." *University of Pennsylvania Law Review* 155, no. 6: 1353–1400.

Allington, David, Sarah Brouillette, David Golumbia. 2016. "Neoliberal Tools (and Archives): A Political History of Digital Humanities." *Los Angeles Review of Books*, May 1. https://lareviewofbooks.org/article/neoliberal-tools-archives-political-history-digital-humanities/.

Angell, James B. 1879. *The Higher Education: A Plea for Making It Accessible to All.* Ann Arbor: University of Michigan Board of Regents.

Angner, Erik. 2011. "The Evolution of Eupathics: The Historical Roots of Subjective Measures of Wellbeing." *International Journal of Wellbeing* 1, no. 1: 4–41.

Angrist, Joshua D., and Jörn-Steffen Pischke. 2008. *Mostly Harmless Econometrics: An Empiricist's Companion*. Princeton, NJ: Princeton University Press.

———. 2010. "The Credibility Revolution in Empirical Economics: How Better Research Design Is Taking the Con out of Econometrics." *Journal of Economic Perspectives* 24: 3–30.

Applebome, Peter. 1992. "The 1992 Campaign: Death Penalty; Arkansas Execution Raises Questions on Governor's Politics." *New York Times*, January 25. https://www.nytimes.com/1992/01/25/us/1992-campaign-death-penalty-arkansas-execution-raises-questions-governor-s.html.

Aristotle. 2000. *Nicomachean Ethics*. Translated by Roger Crisp. Cambridge: Cambridge University Press.

Arrow, Kenneth J. 1962. "The Economic Implications of Learning by Doing." *Review of Economic Studies* 29, no. 3: 155–73.

———. 1973. "Higher Education as a Filter." *Journal of Public Economics* 2, no. 3: 193–216.

Arum, Richard, and Josipa Roksa. 2011. *Academically Adrift: Limited Learning on College Campuses*. Chicago: University of Chicago Press.

Ashcroft, Michael. 2014. "How Scotland Voted, and Why." Lord Ashcroft Polls, September 19. https://lordashcroftpolls.com/2014/09/scotland-voted/.

Associated Press. 2021. "'Do You Miss Me Yet?' At CPAC, Trump Repeats Election Lies, Says He Won't Start Third Party." *MarketWatch*, February 28, 2021. https://www.marketwatch.com/story/do-you-miss-me-yet-trump-takes-center-stage-at-cpac-says-he-wont-start-third-party-01614552672.

Austen, Jane. (1816) 2003. *Emma*. Edited by James Kinsley and Adela Pinch. New York: Oxford University Press.

Azariadis, Costas, and Allan Drazen. 1990. "Threshold Externalities in Economic Development." *Quarterly Journal of Economics* 105, no. 2: 501–26.

"Bachelor's Degrees in the Humanities." 2021. Humanities Indicators. https://www.amacad.org/humanities-indicators/higher-education/bachelors-degrees-humanities.

Badano, Gabriele, Stephen John, and Trenholme Junghans. 2017. "NICE's Cost-Effectiveness Threshold, or: How We Learned to Stop Worrying and (Almost) Love the £20,000–£30,000/QALY Figure." In *Measurement in Medicine: Philosophical Essays on Assessment and Evaluation*, edited by Leah McClimans, 151–69. Lanham, MA: Rowman & Littlefield.

Barabak, Mark Z., and Nigel Duara. 2016. "'We're Called Redneck, Ignorant, Racist. That's Not True': Trump Supporters Explain Why They Voted for Him." *Los Angeles Times*, November 13. http://www.latimes.com/politics/la-na-pol-donald-trump-american-voices-20161113-story.html.

Batty, David. 2003. "Q+A: Five-a-Day Campaign." *The Guardian*, January 20.

Baumol, William J., and Alan S. Blinder. 2007. *Macroeconomics: Principles and Policy*. 10th rev. ed. Cincinnati, OH: South-Western College Publishing.

BBC Staff. 2016. "US Election: Full Transcript of Donald Trump's Obscene Videotape." *BBC News*, October 9. https://www.bbc.com/news/election-us-2016-37595321.

Beauchamp, Tom L., and James F. Childress. 2001. *Principles of Biomedical Ethics*. 5th ed. New York: Oxford University Press.

Becker, Gary S. 1993. *Human Capital: A Theoretical and Empirical Analysis, with Special Reference to Education*. 3rd. ed. Chicago: University of Chicago Press.

Bedard, Kelly. 2001. "Human Capital versus Signaling Models: University Access and High School Dropouts." *Journal of Political Economy* 109, no. 4: 749–75.

Bender, Emily M., Timnit Gebru, et al. 2021. "On the Dangers of Stochastic Parrots: Can Language Models Be Too Big?" In *FAccT '21: Proceedings of the 2021 ACM Conference on Fairness, Accountability, and Transparency*. New York: Association for Computing Machinery, 610–23. https://doi.org/10.1145/3442188.3445922.

Benjamin, Ruha. 2019. *Race after Technology: Abolitionist Tools for the New Jim Code*. Cambridge: Polity Press.

Bennett, Michael, and Jacqueline Brady. 2014. "A Radical Critique of the Learning Outcomes Assessment Movement." *Radical Teacher* 100 (Fall): 34–47.

Bevan, Gwyn, and Christopher Hood. 2006. "What's Measured Is What Matters: Targets and Gaming in the English Public Health Care System." *Public Administration* 84, no. 3: 517–38.

Best, Joel. 2004. *More Damned Lies and Statistics: How Numbers Confuse Public Issues*. Berkeley: University of California Press.

———. 2012. *Damned Lies and Statistics: Untangling Numbers from the Media, Politicians, and Activists*. Updated ed. Berkeley: University of California Press.

Betz, Gregor. 2013. "In Defence of the Value Free Ideal." *European Journal for Philosophy of Science* 3, no. 2: 207–20.

Bickerton, Christopher J., and Carlo Invernizzi Accetti. 2017. "Populism and Technocracy." In *The Oxford Handbook of Populism*, edited by Cristóbal Kaltwasser, Paul Taggart, Paulina Ochoa Espejo, and Pierre Ostiguy, 326–41. Oxford: Oxford University Press.

———. 2018. "'Techno-Populism' as a New Party Family: The Case of the Five Star Movement and Podemos." *Contemporary Italian Politics* 10, no. 2: 132–50.

Biddle, Justin. 2013. "State of the Field: Transient Underdetermination and Values in Science." *Studies in History and Philosophy of Science Part A* 44, no. 1: 124–33.

Biden, Joe (@JoeBiden). 2020. "I believe in science. Donald Trump doesn't. It's that simple, folks." Twitter, October 29. https://twitter.com/joebiden/status/13216064 23495823361?lang=en.

Blankenburg, Wolfgang. 1971. *Der Verlust der Naturlichen Selbstverstaendlichkeit: Ein Beitrag zur Psychopathologie Symtomarmer Schizophrenia*. Stuttgart: Ferdinand Enke.

Blastland, Michael, and Andrew W. Dilnot. 2007. *The Tiger That Isn't: Seeing through a World of Numbers*. London: Profile Books.

Bloomberg Politics. 2016. "Trump to Clinton: 'You Do Have Experience, but It's Bad Experience.'" October 19. YouTube video, 2:45. https://www.youtube.com/watch?v=4XYSxKAmG64.

Bode, Katherine. 2020. "Why You Can't Model away Bias." *Modern Language Quarterly* 81, no. 1: 95–124.

Bogen, James, and James Woodward. 1988. "Saving the Phenomena." *The Philosophical Review* 97, no. 3: 303–52.

Boswell, Christina. 2009. *The Political Uses of Expert Knowledge: Immigration Policy and Social Research*. Cambridge: Cambridge University Press.

———. 2018. *Manufacturing Political Trust: Targets and Performance Measurement in Public Policy*. Cambridge: Cambridge University Press.

Bourdieu, Pierre. 1990. *The Logic of Practice*. Stanford, CA: Stanford University Press.

Bowker, Geoffrey C., and Susan Leigh Star. 2000. *Sorting Things Out: Classification and Its Consequences*. Cambridge, MA: MIT Press.

Boxill, Bernard. 2017. "Kantian Racism and Kantian Teleology." In *The Oxford Handbook of Philosophy and Race*, edited by Naomi Zack, 44–53. Oxford: Oxford University Press.

Boyle, David. 2000. *The Tyranny of Numbers: Why Counting Can't Make Us Happy*. London: HarperCollins.

Brennan, Jason. 2016. *Against Democracy*. Princeton, NJ: Princeton University Press.

Brennan, Timothy. 2017. "The Digital-Humanities Bust: After a Decade of Investment and Hype, What Has the Field Accomplished? Not Much." *Chronicle of Higher Education*, October 15.

Brenner, Marie. 1990. "After the Gold Rush." *Vanity Fair*, September. https://archive .vanityfair.com/article/1990/9/after-the-gold-rush.

Brooks, Cleanth. 1979. "Irony as a Principle of Structure." In *Critical Theory since Plato*, rev. ed., edited by Hazard Adams. New York: Harcourt Brace Jovanovich.

Brown, Philip, Hugh Lauder, and Sin Yi Cheung. 2020. *The Death of Human Capital: Its Failed Promise and How to Renew It in an Age of Disruption*. Oxford: Oxford University Press.

Brown, Wendy. 2015. *Undoing the Demos: Neoliberalism's Stealth Revolution*. New York: Zone Books.

Bruno, Isabelle, Emmanuel Didier, and Tommaso Vitale. 2014. "Statactivism: Forms of Action between Disclosure and Affirmation." *Partecipazione e conflitto: The Open Journal of Sociopolitical Studies* 7, no. 2: 198–220.

Burnham, Peter. 2001. "New Labour and the Politics of Depoliticisation." *British Journal of Politics and International Relations* 3, no. 2: 127–49.

Butler, Judith. 1990. *Gender Trouble: Feminism and the Subversion of Identity*. Chicago: University of Chicago Press.

———. 2009. "Critique, Dissent, Disciplinarity." *Critical Inquiry* 35, no. 4: 787–95.

Buttigieg, Joseph A. 1995. "Gramsci on Civil Society." *Boundary 2* 22, no. 3: 1–32.

Cameron, A. Colin, and Pravin K. Trivedi. 2005. *Microeconometrics: Methods and Applications*. New York: Cambridge University Press.

Campbell, Corbin M. 2015. "Serving a Different Master: Assessing College Educational Quality for the Public." *Higher Education: Handbook of Theory and Research* 30: 525–79.

Cantoni, Davide, Yuyu Chen, David Y. Yang, Noam Yuchtman, and Y. Jane Zhang. 2017. "Curriculum and Ideology." *Journal of Political Economy* 125, no. 2: 338–92.

Capacci, Sara, Mario Mazzocchi, Bhavani Shankar, José Brambila Macias, Wim Verbeke, Federico J. Pérez-Cueto, Agnieszka Kozioł-Kozakowska, Beata Piórecka, Barbara Niedzwiedzka, Dina D'Addesa, and Anna Saba. 2012. "Policies to Promote Healthy Eating in Europe: A Structured Review of Policies and Their Effectiveness." *Nutrition Reviews* 70, no. 3: 188–200.

Caplan, Bryan. 2018. *The Case against Education: Why the Education System Is a Waste of Time and Money*. Princeton, NJ: Princeton University Press.

Caputo, John D. 2000. *More Radical Hermeneutics: On Not Knowing Who We Are*. Bloomington: Indiana University Press.

Carnes, Nicholas, and Noam Lupu. 2017. "It's Time to Bust the Myth: Most Trump Voters Were Not Working Class." *Washington Post*, June 5. https://www.washing

tonpost.com/news/monkey-cage/wp/2017/06/05/its-time-to-bust-the-myth-most
-trump-voters-were-not-working-class/.

Carnevale, Anthony P., and Jeff Strohl. 2013. *Separate & Unequal: How Higher Education Reinforces the Intergenerational Reproduction of White Racial Privilege.* Washington, DC: Georgetown University Center for Education and the Work Force. https://1gyhoq479ufd3yna29x7ubjn-wpengine.netdna-ssl.com/wp-content /uploads/SeparateUnequal.FR_.pdf.

Carpenter, Daniel P. 2001. *The Forging of Bureaucratic Autonomy: Reputations, Networks, and Policy Innovation in Executive Agencies, 1862–1928.* Princeton, NJ: Princeton University Press.

Carson, Thomas L. 2010. *Lying and Deception: Theory and Practice.* Oxford: Oxford University Press.

Cavell, Stanley. 1969a. "The Avoidance of Love: A Reading of *King Lear.*" In *Must We Mean What We Say? A Book of Essays.* Cambridge: Cambridge University Press, 267–353.

———. 1969b. *Must We Mean What We Say? A Book of Essays.* New York: Cambridge University Press.

———. 1979. *The Claim of Reason: Wittgenstein, Skepticism, Morality, and Tragedy.* New York: Oxford University Press.

———. 1981. *The Senses of Walden.* Expanded edition. San Francisco, CA: North-point Press.

Centeno, Miguel Angel. 1993. "The New Leviathan: The Dynamics and Limits of Technocracy." *Theory and Society* 22, no. 3: 307–35.

Centre for Economic Performance. 2016. "Session 2: Why Should Policy Makers Care about People's Wellbeing?" December 20. YouTube video, 1:26:31. https:// youtu.be/yubTfLVFzzw.

Chakrabarty, Dipesh. 2000. *Provincializing Europe.* Princeton, NJ: Princeton University Press.

Chandler, James. 1998. *England in 1819: The Politics of Literary Culture and the Case of Romantic Historicism.* Chicago: University of Chicago Press.

Chatterjee, Elizabeth, and Greg Lusk. 2018. "Deliberative Doubt: Public Ignorance, Epistemic Gatekeeping, and the Problem of Trust." Unpublished manuscript.

Chetty, Raj, John N. Friedman, Emmanuel Saez, Nicholas Turner, and Danny Yagan. 2020. "Income Segregation and Intergenerational Mobility across Colleges in the United States." *Quarterly Journal of Economics* 135, no. 3: 1567–633.

Cheung, Felix, and Richard E. Lucas. 2014. "Assessing the Validity of Single-Item Life Satisfaction Measures: Results from Three Large Samples." *Quality of Life Research: An International Journal of Quality of Life Aspects of Treatment, Care and Rehabilitation* 2320: 2809–18. http://doi.org/10.1007/s11136-014-0726-4.

Cheung, Richard Y., Jillian Cohen, and Patricia Illingworth. 2004. "Orphan Drug Policies: Implications for the United States, Canada, and Developing Countries." *Health Law Journal* 12: 183–200.

Christidis, Nikolaos, Peter A. Stott, Adam A. Scaife, Alberto Arribas, Gareth S. Jones, Dan Copsey, Jeff R. Knight, and Warren J. Tennant. 2013. "A New HadGEM3-A-Based System for Attribution of Weather- and Climate-Related Extreme Events." *Journal of Climate* 26, no. 9: 2756–83. https://doi.org/10.1175 /JCLI-D-12-00169.1.

Chwang, Eric. 2016. "Consent's Been Framed: When Framing Effects Invalidate Consent and How to Validate It Again." *Journal of Applied Philosophy* 33, no. 3: 270–85.

Clark, Andrew, Sarah Flèche, Richard Layard, Nattavudh Powdthavee, and George Ward. 2018. *The Origins of Happiness: The Science of Well-Being over the Life Course.* Princeton, NJ: Princeton University Press.

Clark, Andrew, Richard Layard, and Claudia Senik. 2012. "The Causes of Happiness and Misery." In *World Happiness Report*, edited by John Helliwell, Richard Layard, and Jeffrey Sachs, 59–89. New York: Earth Institute, Columbia University. https://worldhappiness.report/ed/2012/.

Cleland, Carol E. 2001. "Historical Science, Experimental Science, and the Scientific Method." *Geology* 29, no. 1: 987–90.

Clement, Scott. 2017. "Discrimination against Whites Was a Core Concern of Trump's Base." *Washington Post*, August 2, 2017. https://www.washingtonpost.com/news/the-fix/wp/2017/08/02/discrimination-against-whites-was-a-core-concern-of-trumps-base/.

Climate Analytics. n.d. "Global Warming Reaches 1°C above Preindustrial, Warmest in More than 1,000 Years." Accessed June 5, 2019. https://climateanalytics.org/briefings/global-warming-reaches-1c-above-preindustrial-warmest-in-more-than-11000-years/.

Clinton, Bill. 1996. "Text of President Clinton's Announcement on Welfare Legislation." *New York Times*, August 1. https://www.nytimes.com/1996/08/01/us/text-of-president-clinton-s-announcement-on-welfare-legislation.html.

Clots-Figueras, Irma, and Paolo Masella. 2013. "Education, Language and Identity." *The Economic Journal* 123, no. 570: F332–F357.

Coates, Ta-Nehisi. 2015. *Between the World and Me.* New York: Random House.

Cohen, Gerald A. 1993. "Equality of What? On Welfare, Goods, and Capabilities." In *The Quality of Life*, edited by Martha Nussbaum and Amartya Sen, 9–29. Oxford: Clarendon Press.

Colander, David C. 2008. *Economics.* New York: McGraw-Hill/Irwin.

Collier, David. 2011. "Understanding Process Tracing." *PS: Political Science & Politics* 44, no. 4: 823–30.

Confessore, Nicholas, and Nate Cohn. 2016. "Donald Trump's Victory Was Built on Unique Coalition of White Voters." *New York Times*, November 9. http://www.nytimes.com/2016/11/10/us/politics/donald-trump-voters.html.

Cook, John, Dana Nuccitelli, Sarah A. Green, Mark Richardson, Bärbel Winkler, Rob Painting, Robert Way, Peter Jacobs, and Andrew Skuce. 2013. "Quantifying the Consensus on Anthropogenic Global Warming in the Scientific Literature." *Environmental Research Letters* 8, no. 2: 024024. https://doi.org/10.1088/1748-9326/8/2/024024.

Covance. n.d. Phase IV Studies. Labcorp Drug Development. Accessed October 12, 2018. https://www.covance.com/industry-solutions/drug-development/by-phase/phase-iv-solutions.html.

Cox, Daniel, Rachel Lienesch, and Robert P. Jones. 2017. "Beyond Economics: Fears of Cultural Displacement Pushed the White Working Class to Trump." *PRRI/The Atlantic Report*, May 9. https://www.prri.org/research/white-working-class-attitudes-economy-trade-immigration-election-donald-trump/.

Craig, Maureen A., and Jennifer A. Richeson. 2014. "On the Precipice of a 'Majority-Minority' America: Perceived Status Threat from the Racial Demographic Shift Affects White Americans' Political Ideology." *Psychological Science* 25, no. 6: 1189–97.

Creager, Angela N. H., Elizabeth Lunbeck, and M. Norton Wise, eds. 2007. *Science without Laws: Model Systems, Cases, Exemplary Narratives.* Durham, NC: Duke University Press.

Cronon, William. 1991. *Nature's Metropolis: Chicago and the Great West.* New York: W. W. Norton.

Crosby, Alfred W. 1997. *The Measure of Reality: Quantification in Western Europe, 1250–1600.* Cambridge: Cambridge University Press.

Crouch, Colin. 2004. *Post-Democracy.* Cambridge: Polity Press.

Da, Nan Z. 2019a. "The Computational Case against Computational Literary Studies." *Critical Inquiry* 45, no. 3: 601–39.

———. 2019b. "The Digital Humanities Debacle: Computational Methods Repeatedly Come Up Short." *Chronicle of Higher Education*, March 27. https://www .chronicle.com/article/The-Digital-Humanities-Debacle/245986.

Daniels, Norman. 2008. *Just Health: Meeting Health Needs Fairly.* Cambridge: Cambridge University Press.

Dargent, Eduardo. 2015. *Technocracy and Democracy in Latin America: The Experts Running Government.* New York: Cambridge University Press.

Daston, Lorraine. 1992. "Objectivity and the Escape from Perspective." *Social Studies of Science* 22, no. 4: 597–618.

———. 1995. "The Moral Economy of Science." *Osiris* 10: 2–24.

———. 2016. "History of Science without *Structure*." In *Kuhn's "Structure of Scientific Revolutions" at Fifty: Reflections on a Science Classic*, edited by Robert J. Richards and Lorraine Daston, 115–32. Chicago: University of Chicago Press.

Daston, Lorraine, and Peter Galison. 2010. *Objectivity.* New York: Zone Books.

Davidson, Russell, and James G. MacKinnon. 1993. *Estimation and Inference in Econometrics.* New York: Oxford University Press.

Davies, William. 2015. *The Happiness Industry: How the Government and Big Business Sold Us Well-Being.* London: Verso.

———. 2017. "How Statistics Lost Their Power—and Why We Should Fear What Comes Next." *The Guardian*, January 19. https://www.theguardian.com/politics /2017/jan/19/crisis-of-statistics-big-data-democracy.

———. 2018. *Nervous States: Democracy and the Decline of Reason.* New York: W. W. Norton.

Davis, Aeron. 2017. "The New Professional Econocracy and the Maintenance of Elite Power." *Political Studies* 65, no. 3: 594–610.

Davis, Courtney, Joel Lexchin, Tom Jefferson, Peter Gøtzsche, and Martin McKee. 2016. "'Adaptive Pathways' to Drug Authorisation: Adapting to Industry?" *British Medical Journal, Clinical Research Edition* 354: i4437.

Davis, Mike. 2017. "The Great God Trump and the White Working Class." *Jacobin Magazine*, February 7. https://www.jacobinmag.com/2017/02/the-great-god -trump-and-the-white-working-class/.

Deacon, David, and Wendy Monk. 2006 "'New Managerialism' in the News: Media Coverage of Quangos in Britain." *Journal of Public Affairs* 1, no. 2: 153–66.

Dean, John W., and Bob Altemeyer. 2020. *Authoritarian Nightmare*. New York: Melville House.

Deaton, Angus. 2016. "Measuring and Understanding Behavior, Welfare, and Poverty." *American Economic Review* 106, no. 6: 1221–43.

Deaton, Angus, and Arthur A. Stone. 2016. "Response to Lucas, Oishi, and Diener." *Oxford Economic Papers* 68, no. 4: 877–78.

Dee, Thomas S. 2004. "Are There Civic Returns to Education?" *Journal of Public Economics* 88: 1697–720.

De Man, Paul. 1986. *Resistance to Theory*. Minneapolis: University of Minnesota Press.

De Neve, Jan-Emmanuel, Andrew E. Clark, Christian Krekel, Richard Layard, and Gus O'Donnell. 2020. "Taking a Wellbeing Years Approach to Policy Choice." *British Medical Journal* 371: 3853.

Desrosières, Alain. 1998. *The Politics of Large Numbers: A History of Statistical Reasoning*. Translated by Camille Naish. Cambridge, MA: Harvard University Press.

———. 2015. "Retroaction: How Indicators Feed Back onto Quantified Actors." In *The World of Indicators: The Making of Governmental Knowledge through Quantification*, edited by Richard Rottenburg, Sally E. Merry, Sung-Joon Park, and Johanna Mugler, 329–53. Cambridge: Cambridge University Press.

Devega, Chauncey. 2017. "Scholar Justin Gest on the White Working Class: 'They Feel They Are Being Punished' for the Past." *Salon*, September 29. https://www.salon.com/2017/09/29/scholar-justin-gest-on-the-white-working-class-they-feel-they-are-being-punished-for-the-past/.

Dickson, Matt, and Colm Harmon. 2011. "Economic Returns to Education: What We Know, What We Don't Know, and Where We Are Going—Some Brief Pointers." *Economics of Education Review* 30, no. 6: 1118–22.

Diener, Ed, Robert Emmons, Randy Larsen, and Sharon Griffin. 1985. "The Satisfaction with Life Scale." *Journal of Personality Assessment* 49, no. 1: 71–75.

Diener, Ed, Ronald Inglehart, and Louis Tay. 2013. "Theory and Validity of Life Satisfaction Scales." *Social Indicators Research* 112, no. 3: 497–527.

Diener, Ed, Richard Lucas, Ulrich Schimmack, and John Helliwell. 2008. *Well-Being for Public Policy*. New York: Oxford University Press.

Diener, Ed, and Martin E. P. Seligman. 2004. "Beyond Money: Toward an Economy of Well-Being." *Psychological Science in the Public Interest* 5, no. 1: 1–31.

Diener, Ed, and Eunkook M. Suh, eds. 2003. *Culture and Subjective Well-Being*. Cambridge, MA: MIT Press.

D'Ignazio, Catherine, and Lauren Klein. 2020. *Data Feminism*. Cambridge, MA: MIT Press,

Douglas, Heather. 2009. *Science, Policy, and the Value-Free Ideal*. Pittsburgh: University of Pittsburgh Press.

Drucker, Johanna. 2014. *Graphesis: Visual Forms of Knowledge Production*. Cambridge, MA: Harvard University Press.

———. 2020. *Visualization and Interpretation: Humanistic Approaches to Display*. Cambridge, MA: MIT Press.

Du Bois, W. E. B. 1903. *The Souls of Black Folk*. Chicago: A. C. McClurg.

Duncan, Greg J., and Saul D. Hoffman. 1981. "The Incidence and Wage Effects of Overeducation." *Economics of Education Review* 1, no. 1: 75–86.

Earle, Joe, Cahal Moran, and Zach Ward-Perkins. 2017. *The Econocracy: The Perils of Leaving Economics to the Experts*. Manchester: Manchester University Press.

Easterlin, Richard A. 1974. "Does Economic Growth Improve the Human Lot? Some Empirical Evidence." In *Nations and Households in Economic Growth: Essays in Honor of Moses Abramovitz*, edited by Paul A. David and Melvin W. Reder, 89–125. New York: Academic Press.

Eatwell, Roger, and Matthew J. Goodwin. 2018. *National Populism: The Revolt against Liberal Democracy*. London: Pelican.

Eden, Jonathan M., Klaus Wolter, Friederike E. L. Otto, and Geert Jan van Olden-borgh. 2016. "Multi-Method Attribution Analysis of Extreme Precipitation in Boulder, Colorado." *Environmental Research Letters* 11, no. 12: 124009.

Edsall, Thomas B. 2017. "The Democratic Party Is in Worse Shape than You Thought." *New York Times*, June 8. https://www.nytimes.com/2017/06/08/opinion/the-democratic-party-is-in-worse-shape-than-you-thought.html.

———. 2018. "The Contract with Authoritarianism." *New York Times*, April 5. https://www.nytimes.com/2018/04/05/opinion/trump-authoritarianism-republicans-contract.html.

Emery, Kim. 2008. "Outcomes Assessment and Standardization: A Queer Critique." *Profession*: 255–59. https://www.jstor.org/stable/25595900.

Ender, Evelyne, and Deidre Lynch, eds. 2018. "Cultures of Reading." Special issue, *PMLA* 133, no. 5.

Entous, Adam, and Ronan Farrow. 2018. "The Conspiracy Memo about Obama Aides That Circulated in the Trump White House." *New Yorker*, August 23. https://www.newyorker.com/news/news-desk/the-conspiracy-memo-aimed-at-obama-aides-that-circulated-in-the-trump-white-house.

Espeland, Wendy Nelson. 2015. "Narrating Numbers." In *The World of Indicators: The Making of Governmental Knowledge through Quantification*, edited by Richard Rottenburg, Sally E. Merry, Sung-Joon Park, and Johanna Mugler, 56–75. Cambridge: Cambridge University Press.

Espeland, Wendy Nelson, and Michael Sauder. 2012. "The Dynamism of Indicators." In *Governance by Indicators: Global Power through Quantification and Rankings*, edited by Kevin E. Davis, Angelina Fisher, Benedict Kingsbury, and Sally Engle Merry, 86–109. Oxford: Oxford University Press.

———. 2016. *Engines of Anxiety: Academic Rankings, Reputation, and Accountability*. New York: Russell Sage Foundation.

Espeland, Wendy Nelson, and Mitchell Stevens. 1998. "Commensuration as a Social Process." *Annual Review of Sociology* 24, no. 1: 313–43.

Estlund, David M. 2008. *Democratic Authority: A Philosophical Framework*. Princeton, NJ: Princeton University Press.

Eubanks, David. 2017. "A Guide for the Perplexed." *Intersection* (Fall): 4–13. https://www.aalhe.org/assets/docs/AAHLE_Fall_2017_Intersection.pdf.

Eubanks, Virginia. 2018. *Automating Inequality: How High-Tech Tools Profile, Police, and Punish the Poor*. New York: St. Martin's Press.

European Patients' Academy on Therapeutic Innovation (EUPATI). n.d. "Health Technology Assessment Process: Fundamentals." Accessed July 26, 2021. https://toolbox.eupati.eu/resources/health-technology-assessment-process-fundamentals/.

EURORDIS (Rare Diseases Europe). 2020. "What Is a Rare Disease?" EURORDIS, July 21. https://www.eurordis.org/content/what-rare-disease.

———. n.d. "EURORDIS Summer School." Accessed October 6, 2018. https://openacademy.eurordis.org/summerschool/.

EvaluatePharma. 2014. *Orphan Drug Report 2014*. London: Evaluate. https://info.
evaluategroup.com/rs/evaluatepharmaltd/images/2014OD.pdf.

―――. 2017a. "Median Cost per Patient for Orphan Drugs Is 5.5 Times Higher
than Non-Orphan Drugs." February 28. https://www.evaluate.com/about/press
-releases/median-cost-patient-orphan-drugs-55-times-higher-non-orphan-drugs.

―――. 2017b. *Orphan Drug Report 2017*. London: Evaluate. http://info.evaluate
group.com/rs/607-YGS-364/images/EPOD17.pdf.

―――. 2018. *Orphan Drug Report 2018*. London: Evaluate. https://www.evaluate.com
/sites/default/files/media/download-files/OD18.pdf.

Ewell, Peter T. 2008. "Assessment and Accountability in America Today: Background and Context." *New Directions for Institutional Research* S1 (Fall): 7–17.

Express KCS. 2015. "We Need to Measure Well-Being to Reverse Britain's Social
Decline." City A.M., October 29. https://www.cityam.com/we-need-measure
-well-being-reverse-britain-s-social-decline/.

Fabian, Mark. 2018. "Racing from Subjective Well-Being to Public Policy: A Review
of *The Origins of Happiness*." *Journal of Happiness Studies* 20, no. 6: 2011–26.

Felski, Rita. 2020. *Hooked: Art and Attachment*. Chicago: University of Chicago Press.

Ferrer-i-Carbonell, Ada, and Paul Frijters. 2004. "How Important Is Methodology
for the Estimates of the Determinants of Happiness?" *The Economic Journal* 114,
no. 497: 641–59.

Fichte, Johann G. 1988. *Early Philosophical Writings*. Translated and edited by Daniel
Breazeale. Ithaca, NY: Cornell University Press.

Finlayson, Alan. 2017. "Brexitism." *London Review of Books* 39, no. 10. https://www
.lrb.co.uk/the-paper/v39/n10/alan-finlayson/brexitism.

Fioramonti, Lorenzo. 2014a. *How Numbers Rule the World: The Use and Abuse of
Statistics in Global Politics*. London: Zed Books.

―――. 2014b. "The Politics of Numbers in the Age of Austerity." OpenDemocracy,
January 25. https://www.opendemocracy.net/can-europe-make-it/lorenzo
-fioramonti/politics-of-numbers-in-age-of-austerity.

Fish, Stanley. 1982. *Is There a Text in This Class? The Authority of Interpretive
Communities*. Cambridge, MA: Harvard University Press.

Flanders, Julia. 2005. "Detailism, Digital Texts, and the Problem of Pedantry." *TEXT
Technology* 2: 41–70.

Fleming, Paul. 2012. "On the Edge of Non-Contingency: Anecdotes and the Life-
world." *Telos* 158: 21–35.

Flinders, Matthew. 2012. *Defending Politics: Why Democracy Matters in the Twenty-
First Century*. Oxford: Oxford University Press.

Flinders, Matthew, and Jim Buller. 2006. "Depoliticisation: Principles, Tactics
and Tools." *British Politics* 1, no. 3: 293–318.

Forrester, John. 1996. "If *p*, Then What? Thinking in Cases." *History of the Human
Sciences* 9, no. 3: 1–25.

―――. 1997. *Dispatches from the Freud Wars: Psychoanalysis and Its Passions*.
Cambridge, MA: Harvard University Press.

―――. 2016. *Thinking in Cases*. Malden, MA: Polity Press.

Foucault, Michel. 1980. "The Confession of the Flesh: An Interview." In *Power/
Knowledge: Selected Interviews and Other Writings, 1972–1977*, edited by Colin
Gordon, 194–228. New York: Pantheon Books.

Fourcade, Marion. 2016. "Ordinalization." *Sociological Theory* 34, no. 3: 175–95.

Freeman, Richard. 1976. *The Overeducated American*. New York: Academic Press.

———. 2006. "The Great Doubling: The Challenge of the New Global Labor Market." UC Berkeley Econometrics Laboratory. https://eml.berkeley.edu/~webfac /eichengreen/e183_sp07/great_doub.pdf.

Fretz, Thomas A., ed. 2008. "The Morrill Land Grant Act of 1862 and the Changing of Higher Education in America." http://escop.info/wp-content/uploads/2017/04 /Morrill-Land-Grant-Act-and-Impacts.pdf.

Freud, Sigmund. (1915) 1966. "Repression." In *The Standard Edition of the Complete Psychological Works of Sigmund Freud*, translated and edited by James Strachey, Anna Freud, Alix Strachey, and Alan Tyson. Vol. 14, *On the History of the Psycho-Analytic Movement, Papers on Meta-Psychology and Other Works*, 141–58. London: Hogarth Press.

Friedman, Milton. 2002. *Capitalism and Freedom*. Fortieth anniversary edition. New Brunswick, NJ: Rutgers University Press.

Friedman, Willa, Michael Kremer, Edward Miguel, and Rebecca Thornton. 2016. "Education as Liberation?" *Economica* 83, no. 329: 1–30.

Frigg, Roman, Seamus Bradley, Hailiang Du, and Leonard A. Smith. 2014. "Laplace's Demon and the Adventures of His Apprentices." *Philosophy of Science* 81, no. 1: 31–59.

Frijters, Paul, Andrew E. Clark, Christian Krekel, and Richard Layard. 2020. "A Happy Choice: Wellbeing as the Goal of Government." *Behavioural Public Policy* 4, no. 2: 126–65.

Frye, Northrop. 1968. *A Study of English Romanticism*. New York: Random House.

Fujiwara, Daniel, and Ross Campbell. 2011. *Valuation Techniques for Social Cost-Benefit Analysis: Stated Preference, Revealed Preference and Subjective Well-Being Approaches; A Discussion of the Current Issues*. London: Her Majesty's Treasury.

Gadamer, Hans-Georg. 1988. "On the Circle of Understanding." In Hans-Georg Gadamer, Ernst Konrad Specht, and Wolfgang Stegmüller, *Hermeneutics versus Science? Three German Views*, translated and edited by John M. Connolly and Thomas Keutner, 68–78. Notre Dame, IN: University of Notre Dame Press.

Galperin, William, ed. 2000. *Re-Reading Box Hill: Reading the Practice of Reading; Everyday Life*. Boulder, CO: Romantic Circles Praxis. https://www.rc.umd.edu /praxis/boxhill/index.html.

Gavin, Michael. 2020. "Is There a Text in My Data? (Part 1): On Counting Words." *Journal of Cultural Analytics* 1, no. 1. https://doi.org/10.22148/001c.11830.

Gebru, Timnit, et al. 2020. "Datasheets for Datasets." March 19. https://arxiv.org /abs/1803.09010.

Geertz, Clifford. 2007. "'To Exist Is to Have Confidence in One's Way of Being': Rituals as Model Systems." In *Science without Laws: Model Systems, Cases, Exemplary Narratives*, edited by Angela N. H. Creager, Elizabeth Lunbeck, and M. Norton Wise, 212–24. Durham, NC: Duke University Press.

Gere, Cathy. 2017. *Pain, Pleasure, and the Greater Good: From the Panopticon to the Skinner Box and Beyond*. Chicago: University of Chicago Press.

Gigerenzer, Gerd, and Odette Wegwarth. 2013. "Five Year Survival Rates Can Mislead." *British Medical Journal* 346: f548. https://www.bmj.com/content/346/bmj.f548.

Gillis, Justin. 2011. "Study Links Rise in Rain and Snow to Human Actions." *New York Times*, February 17. https://www.nytimes.com/2011/02/17/science/earth /17extreme.html.

Ginzburg, Carlo. 2007. "Latitude, Slaves, and the Bible: An Experiment in Microhistory." In *Science without Laws: Model Systems, Cases, Exemplary Narratives*, edited by Angela N. H. Creager, Elizabeth Lunbeck, and M. Norton Wise, 243–63. Durham, NC: Duke University Press.

Gladwell, Malcolm. 2008. *Outliers: The Story of Success*. New York: Little, Brown.

Glaeser, Edward L., Giacomo A. M. Ponzetto, and Andrei Shleifer. 2007. "Why Does Democracy Need Education?" *Journal of Economic Growth* 12: 77–99.

Glaeser, Edward L., and Albert Saiz. 2003. "The Rise of the Skilled City." National Bureau of Economic Research Working Paper 10191. https://www.nber.org /papers/w10191.

Global Genes. n.d. "RARE Facts." Accessed February 27, 2021. https://globalgenes .org/rare-facts/.

Global Strategy Group and Garin Hart Yang. 2017. "Post-Election Research: Persuadable and Drop-Off Voters." Priorities USA. https://www.washingtonpost .com/r/2010-2019/WashingtonPost/2017/05/01/Editorial-Opinion/Graphics /Post-election_Research_Deck.pdf?tid=a_inl.

Goldberger, Arthur Stanley. 1991. *A Course in Econometrics*. Cambridge, MA: Harvard University Press.

Goldrick-Rab, Sara. 2016. *Paying the Price: College Costs, Financial Aid, and the Betrayal of the American Dream*. Chicago: University of Chicago Press.

Golshan, Tara. 2016. "Full Transcript: Hillary Clinton and Donald Trump's Final Presidential Debate." *Vox*, October 19. https://www.vox.com/policy-and-politics /2016/10/19/13336894/third-presidential-debate-live-transcript-clinton-trump.

Gonyea, Don. 2017. "Majority of White Americans Say They Believe Whites Face Discrimination." NPR, October 24. http://www.npr.org/2017/10/24/559604836 /majority-of-white-americans-think-theyre-discriminated-against.

Good, James A. 2014. "The German *Bildung* Tradition." Unpublished manuscript, July 7. https://www.academia.edu/8055120/The_German_Bildung_Tradition.

Goodman, Nelson. (1972) 1992. "Seven Strictures on Similarity." In *How Classification Works: Nelson Goodman among the Social Sciences*, edited by Mary Douglas and David Hull, 13–23. Edinburgh: Edinburgh University Press.

Goodwin, Jonathan. 2006. "Franco Moretti's *Graphs Maps Trees*: A Valve Book Event." *The Valve*, January 2.

Goodwin, Matthew J., and Oliver Heath. 2016. "The 2016 Referendum, Brexit and the Left Behind: An Aggregate-Level Analysis of the Result." *Political Quarterly* 87, no. 3: 323–32.

Google Scholar. 2018. "Top Publications, Categories: Humanities, Literature & Arts." https://scholar.google.com/citations?view_op=top_venues&hl=en&vq=hum.

Gorur, Radhika. 2014. "Towards a Sociology of Measurement in Education." *European Educational Research Journal* 12, no. 1: 58–72.

Gottlieb, Scott. 2018. "FDA's Comprehensive Effort to Advance New Innovations: Initiatives to Modernize for Innovation." US Food and Drug Administration (FDA), August 29. https://www.fda.gov/news-events/fda-voices/fdas-compre hensive-effort-advance-new-innovations-initiatives-modernize-innovation.

Gould, Philip. 1998. *The Unfinished Revolution: How the Modernisers Saved the Labour Party*. London: Little, Brown.

Graeber, David. 2018. *Bullshit Jobs: A Theory*. New York: Simon and Schuster.

Graff, Gerald, and Cathy Birkenstein. 2011. "A Progressive Case for Educational Standardization: How Not to Respond to Calls for Common Standards." In *Literary Study, Measurement, and the Sublime: Disciplinary Assessment*, edited by Donna Heiland and Laura J. Rosethal, 217–26. New York: Teagle Foundation.

Gray, John. 2017. "Post Truth by Matthew D'Ancona and Post-Truth by Evan Davis: Is This Really a New Era of Politics?" *The Guardian*, May 19. https://www.theguardian.com/books/2017/may/19/post-truth-matthew-dancona-evan-davis-reiews.

Green, Erica L., Matt Apuzzo, and Katie Benner. 2018. "Trump Officials Reverse Obama's Policy on Affirmative Action in Schools." *New York Times*, July 4. https://www.nytimes.com/2018/07/03/us/politics/trump-affirmative-action-race-schools.html.

Greenberg, Stanley B. 2017. "The Democrats' 'Working-Class Problem.'" *The American Prospect*, June 1. http://prospect.org/article/democrats%E2%80%99-%E2%80%98working-class-problem%E2%80%99.

Greene, William H. 2008. *Econometric Analysis*. Upper Saddle River, NJ: Pearson Prentice Hall.

Grice, Paul. 1975. "Logic and Conversation." In *Syntax and Semantics 3: Speech Acts*, edited by Peter Cole and Jerry Morgan, 41–58. New York: Academic Press.

Grieve, Richard, Keith Abrams, Karl Claxton, et al. 2016. "Cancer Drugs Fund Requires Further Reform." *British Medical Journal* 354: i5090.

Guillory, John. 2002. "The Sokal Affair and the History of Criticism." *Critical Inquiry* 28, no. 2: 470–508.

Gujarati, Damodar N. 1992. *Essentials of Econometrics*. New York: McGraw-Hill.

Hacking, Ian. 1990. *The Taming of Chance*. Cambridge: Cambridge University Press.

———. 1995. *The Emergence of Probability*. Cambridge: Cambridge University Press.

Haddad, Christian, Haidan Chen, and Herbert Gottweis. 2013. "Unruly Objects: Novel Innovation Paths, and Their Regulatory Challenge." In *The Global Dynamics of Regenerative Medicine*, edited by Andrew Webster, 88–117. London: Palgrave Macmillan.

Hahn, Ulrike, and Michael Ramscar, eds. 2001. *Similarity and Categorization*. Oxford: Oxford University Press.

Hammond, Zaretta L. 2014. *Culturally Responsive Teaching and the Brain: Promoting Authentic Engagement and Rigor among Culturally and Linguistically Diverse Students*. Thousand Oaks, CA: Corwin.

Hampshire, James. 2018. "The Measure of a Nation." *Political Quarterly* 89, no. 3: 370–76.

Hankins, Joseph, and Rihan Yeh. 2016. "To Bind and to Bound: Commensuration across Boundaries." *Anthropological Quarterly* 89, no. 1: 5–30.

Hao, Karen. 2020. "We Read the Paper That Forced Timnit Gebru out of Google: Here's What It Says." *MIT Technology Review*, December 4. https://www.technologyreview.com/2020/12/04/1013294/google-ai-ethics-research-paper-forced-out-timnit-gebru/.

Haraway, Donna. 1990. *Primate Visions: Gender, Race, and Nature in the World of Modern Science*. New York: Routledge.

Harding, Sandra. 2004. "Introduction: Standpoint Theory as a Site of Political, Philosophical, and Scientific debate." In *The Feminist Standpoint Theory Reader: Intellectual and Political Controversies*, edited by Sandra Harding, 1–14. New York: Routledge.

Harney, Stefano, and Fred Moten. 2016. *The Undercommons: Fugitive Planning and Black Study*. London: Minor Compositions.

Harris, John. 2005a. "It's Not NICE to Discriminate." *Journal of Medical Ethics* 31, no. 7: 373–75.

———. 2005b. "Nice and Not So Nice." *Journal of Medical Ethics* 31, no. 12: 685–88.

Harvard University Committee and James Bryant Conant. 1945. *General Education in a Free Society: Report of the Committee*. Cambridge, MA: Harvard University Press.

Hassan, Steven. 2020. *The Cult of Trump*. New York: Simon and Schuster.

Hausman, Daniel M. 2015. *Valuing Health: Well-Being, Freedom, and Suffering*. New York: Oxford University Press.

Hay, Colin. 2007. *Why We Hate Politics*. Cambridge: Polity Press.

Haybron, Daniel M. 2008. *The Pursuit of Unhappiness: The Elusive Psychology of Well-Being*. New York: Oxford University Press.

Heiland, Donna, and Laura J. Rosenthal, eds. 2011. *Literary Study, Measurement, and the Sublime: Disciplinary Assessment*. New York: Teagle Foundation.

Helleiner, Eric. 2014. *The Status Quo Crisis: Global Financial Governance after the 2008 Meltdown*. Oxford: Oxford University Press.

Henry, Charles P. 2017. *Black Studies and the Democratization of Higher Education*. London: Palgrave.

Herder, Matthew. 2013. "When Everyone Is an Orphan: Against Adopting a US-Styled Orphan Drug Policy in Canada." *Accountability in Research* 20, no. 4: 227–69.

Hersh, Eitan. 2015. *Hacking the Electorate: How Campaigns Perceive Voters*. New York: Cambridge University Press.

Hoffman, Carl. 2020. *Liar's Circus*. New York: Custom House.

Hofstadter, Richard. 1962. *Anti-Intellectualism in American Life*. New York: Knopf.

Hogle, Linda F. 2018. "Intersections of Technological and Regulatory Zones in Regenerative Medicine." In *Global Perspectives on Stem Cell Technologies*, edited by Adithya Bharadwaj, 51–84. London: Palgrave Macmillan.

Holmberg, Tora, Nete Schwennesen, and Andrew Webster. 2011. "Bio-Objects and the Bio-Objectification Process." *Croatian Medical Journal* 52, no. 6: 740–42.

Howick, Jeremy. 2011. *The Philosophy of Evidence-Based Medicine*. Oxford: Wiley.

Hsiao, Cheng. 2014. *Analysis of Panel Data*. Cambridge University Press.

Huff, Darrell. 1954. *How to Lie with Statistics*. New York: Norton.

Hulme, Mike. 2014. "Attributing Weather Extremes to 'Climate Change': A Review." *Progress in Physical Geography* 38, no. 4: 499–511.

Hulme, Mike, Saffron J. O'Neill, and Suraje Dessai. 2011. "Is Weather Event Attribution Necessary for Adaptation Funding?" *Science* 334, no. 6057: 764–65.

Huneman, Philippe, Gerard Lambert, and Marc Silberstein. 2015. "Introduction: Surveying the Revival in the Philosophy of Medicine." In *Classification, Disease and Evidence*, edited by Philippe Huneman, Philippe, Gerard Lambert, and Marc Silberstein, vii–xx. New York: Springer.

Huppert, Felicia A., Nicky Baylis, and Barry Keverne. 2005. *The Science of Well-Being*. New York: Oxford University Press.

Ifill, Gwen. 2016. "Questions for President Obama: A Town Hall Special." *PBS NewsHour*, June 1. https://www.pbs.org/newshour/show/questions-for-president -obama-a-town-hall-special.

Imbens, Guido W. 2010. "Better LATE than Nothing: Some Comments on Deaton 2009 and Heckman and Urzua 2009." *Journal of Economic Literature* 48, no. 2: 399–423.

Inglehart, Ronald F., and Pippa Norris. 2016. "Trump, Brexit, and the Rise of Populism: Economic Have-Nots and Cultural Backlash." Harvard Kennedy School Faculty Research Working Paper RWP16-026. https://www.hks.harvard.edu /publications/trump-brexit-and-rise-populism-economic-have-nots-and-cultural -backlash.

Inoue, Asao B. 2015. *Antiracist Writing Assessment Ecologies: Teaching and Assessing Writing for a Socially Just Future*. Fort Collins, CO: WAC Clearinghouse.

Jackson, Jennifer. 1991. "Telling the Truth in Medicine." *Journal of Medical Ethics* 17, no. 1: 5–9

JillD55. 2018. "20 Floors on Flat Ground?" Fitbit Community, August 6, 2018. https:// community.fitbit.com/t5/Fitbit-com-Dashboard/20-floors-on-flat-ground/td -p/2879299.

Jo, Eun Seo, and Timnit Gebru. 2020. "Lessons from the Archive: Strategies for Collecting Sociocultural Data in Machine Learning." In *Conference on Fairness, Accountability, and Transparency (FAT* '20), January 27–30, 2020*, Barcelona, Spain. New York: Association for Computing Machinery, 306–16. https://doi.org /10.1145/3351095.3372829.

Jockers, Matthew. 2011. "The LDA Buffet Is Now Open: Or, Latent Dirichlet Allocation for English Majors." September 29. http://www.matthewjockers.net/2011 /09/29/the-lda-buffet-is-now-open-or-latent-dirichlet-allocation-for-english -majors/.

———. 2013. *Macroanalysis: Digital Methods and Literary History*. Chicago: University of Illinois Press.

John, Stephen. 2015. "The Example of the IPCC Does Not Vindicate the Value Free Ideal: A Reply to Gregor Betz." *European Journal for Philosophy of Science* 5, no. 1: 1–13.

———. 2018. "Epistemic Trust and the Ethics of Science Communication: Against Transparency, Openness, Sincerity and Honesty." *Social Epistemology* 32, no. 2: 75–87.

———. 2021. "Scientific Deceit." *Synthese* 198, no. 1: 373–94.

Johnston, John. 1972. *Econometric Methods*. New York: McGraw-Hill.

Judge, George G., Rufus Carter Hill, William Griffiths, and Tsoung Chao Lee. 1980. *Introduction to the Theory and Practice of Econometrics*. New York: John Wiley and Sons.

Jukola, Saana. 2019. "On the Evidentiary Standards for Nutrition Advice." *Studies in History and Philosophy of Science Part C: Studies in History and Philosophy of Biology and Biomedical Sciences* 73: 1–9.

Kahneman, Daniel. 2011. *Thinking, Fast and Slow*. New York: Farrar, Straus and Giroux.

Kahneman, Daniel, Ed Diener, and Norbert Schwarz, eds. 1999. *Well-Being: The Foundations of Hedonic Psychology*. New York: Russell Sage Foundation.

Kahneman, Daniel, and Alan B. Krueger. 2006. "Developments in the Measurement of Subjective Well-Being." *Journal of Economic Perspectives* 20, no. 1: 3–24.

Kahneman, Daniel, Alan B. Krueger, David A. Schkade, Norbert Schwarz, and Arthur A. Stone. 2004a. "A Survey Method for Characterizing Daily Life Experience: The Day Reconstruction Method." *Science* 306, no. 5702: 1776–80.

———. 2004b. "Toward National Well-Being Accounts." *American Economic Review* 94, no. 2: 429–34.

Kant, Immanuel. 1997. "On a Supposed Right to Lie from Philanthropy." In *Immanuel Kant: Practical Philosophy*, translated by Mary Gregor, 609–15. Cambridge: Cambridge University Press.

Katikireddi, S. Vittal, and Sean A. Valles. 2015. "Coupled Ethical–Epistemic Analysis of Public Health Research and Practice: Categorizing Variables to Improve Population Health and Equity." *American Journal of Public Health* 105, no. 1: e36–e42.

Kelly, Ann H., and Linsey McGoey. 2018. "Facts, Power and Global Evidence: A New Empire of Truth." *Economy and Society* 47, no. 1: 1–26.

Kennedy, Peter E. 2008. *A Guide to Econometrics*. Malden, MA: Wiley-Blackwell.

Kesselheim, Aaron S., Carolyn L. Treasure, and Steven Joffe. 2017. "Biomarker-Defined Subsets of Common Diseases: Policy and Economic Implications of Orphan Drug Act Coverage." *PLOS Medicine* 14, no. 1: e1002190.

Kessler, Glenn. 2021. "In Four Years, President Trump Made 30,573 False or Misleading Claims." *Washington Post*, January 20. https://www.washingtonpost.com /graphics/politics/trump-claims-database/.

Kiker, B. F., Maria C. Santos, and M. Mendes de Oliveira. 1997. "Overeducation and Undereducation: Evidence from Portugal." *Economics of Education Review* 16, no. 2: 111–25.

King, Gary, Robert O. Keohane, and Sidney Verba. 1994. *Designing Social Inquiry: Scientific Inference in Qualitative Research*. Princeton, NJ: Princeton University Press.

Kitcher, Philip. 2003. *Science, Truth and Democracy*. Oxford: Oxford University Press.

Klein, Lauren F. 2016. "Distant Reading after Moretti." Talk given at the MLA Convention, Varieties of Digital Humanities. http://lklein.com/2018/01/distant -reading-after-moretti/.

———. 2019. "What the New Computational Rigor Should Be." *Critical Inquiry: In the Moment*, April 1. In "Computational Literary Studies: Participant Forum Responses." https://critinq.wordpress.com/2019/04/01/computational-literary -studies-participant-forum-responses-5/.

Kramnick, Jonathan. 2011. "Against Literary Darwinism." *Critical Inquiry* 37, no. 2: 315–47.

Krueger, Alan B., and Mikael Lindahl. 2001. "Education for Growth: Why and for Whom?" *Journal of Economic Literature* 39, no. 4: 1101–36.

Krugman, Paul. 2017. "Their Own Private Pyongyang." *New York Times*, June 13. https://krugman.blogs.nytimes.com/2017/06/13/their-own-private-pyongyang/.

Krugman, Paul, and Robin Wells. 2006. *Economics*. New York: Worth.

Kuttner, Robert. 2014. "Obama's Obama: The Contradictions of Cass Sunstein." *Harper's Magazine*, December 1. https://harpers.org/archive/2014/12/obamas-obama/.

Lakoff, George. 1996. *Moral Politics: What Conservatives Know That Liberals Don't*. Chicago: University of Chicago Press.

Lange, Fabian, and Robert Topel. 2006. "The Social Value of Education and Human Capital." In *Handbook of the Economics of Education*, vol. 1, edited by Eric Hanushek and Finis Welch, 459–509. Amsterdam: Elsevier.

Larroulet-Philippi, Cristian. 2021. "On Measurement Scales: Neither Ordinal nor Interval?" Preprint. http://philsci-archive.pitt.edu/id/eprint/19169.

Lauter, David. 1993. "Clinton Withdraws Guinier as Nominee for Civil Rights Job: Justice Department: The President Says He Only Lately Read Her Legal Writings. He Decided She Stood for Principles He Could Not Support in a Divisive Confirmation Battle." *Los Angeles Times*, June 4. https://www.latimes.com/archives/la-xpm-1993-06-04-mn-43290-story.html.

Layard, Richard. 2005. *Happiness: Lessons from a New Science.* London: Penguin.

Lemke, Thomas. 2001. "'The Birth of Bio-Politics': Michel Foucault's Lecture at the Collège de France on Neo-Liberal Governmentality." *Economy and Society* 30, no. 2: 190–207.

Leonard, Thomas C. 2008. Review of Richard H. Thaler and Cass R. Sunstein, *Nudge: Improving Decisions about Health, Wealth, and Happiness. Constitutional Political Economy* 19, no. 4: 356–60.

Lepore, Jill. 2015. "Politics and the New Machine." *New Yorker*, November 8. https://www.newyorker.com/magazine/2015/11/16/politics-and-the-new-machine.

Levy, Ed. 2001. "Quantification, Mandated Science and Judgment." *Studies in History and Philosophy of Science Part A* 32, no. 4: 723–37.

Lichtenberg, Pesach, Uriel Heresco-Levy, and Uriel Nitzan. 2004. "The Ethics of the Placebo in Clinical Practice." *Journal of Medical Ethics* 30, no. 6: 551–55.

Lieberman, Robert C., Suzanne Mettler, Thomas B. Pepinsky, Kenneth M. Roberts, and Richard Valelly. 2017. "Trumpism and American Democracy: History, Comparison, and the Predicament of Liberal Democracy in the United States." *SSRN*, August 31. https://papers.ssrn.com/sol3/papers.cfm?abstract_id=3028990.

Lind, Michael. 2020. *The New Class War: Saving Democracy from the Managerial Elite.* New York: Penguin.

Liu, Alan. 2013. "What Is the Meaning of the Digital Humanities to the Humanities?" *PMLA* 128, no. 2: 409–23.

Lloyd, Elisabeth A., and Naomi Oreskes. 2018. "Climate Change Attribution: When Is It Appropriate to Accept New Methods?" *Earth's Future* 6, no. 3: 311–25.

Lochner, Lance. 2011. "Non-Production Benefits of Education: Crime, Health, and Good Citizenship." National Bureau of Economic Research Working Paper 16722. https://www.nber.org/papers/w16722.

Long, Hoyt, and Richard Jean So. 2016. "Literary Pattern Recognition: Modernism between Close Reading and Machine Learning." *Critical Inquiry* 42, no. 2: 235–67.

Lothrop, George van Ness. 1878. "A Plea for Education as a Public Duty." Ann Arbor: University of Michigan Board of Regents. http://www.umich.edu/~bhlumrec/c/commence/1878-Lothrop.pdf.

Lott, John R. 1999. "Public Schooling, Indoctrination, and Totalitarianism." *Journal of Political Economy* 107, no. S6: S127–S157.

Loughnot, David. 2005. "Potential Interactions of the Orphan Drug Act and Pharmacogenomics: A Flood of Orphan Drugs and Abuses?" *American Journal of Law and Medicine* 31, no. 2–3: 365–80.

Lucas, Richard E., and Nicole M. Lawless 2013. "Does Life Seem Better on a Sunny Day? Examining the Association between Daily Weather Conditions and Life Satisfaction Judgments." *Journal of Personality and Social Psychology* 104, no. 5: 872–84.

Lucas, Richard E., Shigehiro Oishi, and Ed Diener. 2016. "What We Know about Context Effects in Self-Report Surveys of Well-Being: Comment on Deaton and Stone." *Oxford Economic Papers* 68, no. 4: 871–76.

Lucas, Robert E. 1988. "On the Mechanics of Economic Development." *Journal of Monetary Economics* 22: 3–42.

Lupton, Deborah. 2017. *The Quantified Self.* Cambridge: Polity Press.

Lusk, Greg. 2017. "The Social Utility of Event Attribution: Liability, Adaptation, and Justice-Based Loss and Damage." *Climatic Change* 143, no. 1–2: 201–12. https://doi.org/10.1007/s10584-017-1967-3.

Luterbacher, Jürg, Daniel Dietrich, Elena Xoplaki, Martin Grosjean, and Heinz Wanner. 2004. "European Seasonal and Annual Temperature Variability, Trends, and Extremes since 1500." *Science* 303, no. 5663: 1499–1503.

Machin, Stephen, and Anna Vignoles. 2018. *What's the Good of Education? The Economics of Education in the UK.* Princeton, NJ: Princeton University Press.

Maddala, Gangadharrao Soundaryarao. 1977. *Econometrics.* New York: McGraw-Hill.

———. 1992. *Introduction to Econometrics.* New York: Macmillan.

Maher, Paul, and Marlene Haffner. 2006. "Orphan Drug Designation and Pharmacogenomics." *BioDrugs* 20, no. 2: 71–79.

Mair, Peter. 2013. *Ruling the Void: The Hollowing of Western Democracy.* London: Verso.

Mandell, Laura. 2015. *Breaking the Book: Print Humanities in the Digital Age.* Malden, MA: Wiley-Blackwell.

———. 2019. "Gender and Cultural Analytics: Finding or Making Stereotypes?" In *Debates in Digital Humanities,* edited by Matthew K. Gold and Lauren F. Klein, 3–26. Minneapolis: University of Minnesota Press.

Manson, Neil, and Onora O'Neill. 2007. *Rethinking Informed Consent in Bioethics.* Cambridge: Cambridge University Press.

Marginson, Simon. 2016. *Higher Education and the Common Good.* Carlton, Vic.: Melbourne University Press.

Marx, Karl, and Friedrich Engels. 1978. *The Marx-Engels Reader.* Translated by Robert C. Tucker. New York: W. W. Norton.

Mazella, David. 2011. "English Departments, Assessment, and Organizational Learning." In *Literary Study, Measurement, and the Sublime: Disciplinary Assessment,* edited by Donna Heiland and Laura J. Rosenthal, 227–58. New York: Teagle Foundation.

McCabe, Cristopher, Karl Claxton, and Anthony J. Culyer. 2008. "The NICE Cost-Effectiveness Threshold: What It Is and What That Means." *Pharmacoeconomics* 26, no. 9: 733–44.

McDonnell, Duncan, and Marco Valbruzzi. 2014. "Defining and Classifying Technocrat-Led and Technocratic Governments." *European Journal of Political Research* 53, no. 4: 654–71.

McElwee, Sean, and Jason McDaniel. 2017. "Economic Anxiety Didn't Make People Vote Trump, Racism Did." *The Nation,* May 8. https://www.thenation.com/article/economic-anxiety-didnt-make-people-vote-trump-racism-did/.

McGoey, Linsey. 2012. "Strategic Unknowns: Towards a Sociology of Ignorance." *Economy and Society* 41, no. 1: 1–16.

McGuinness, Seamus. 2006. "Overeducation in the Labour Market." *Journal of Economic Surveys* 20, no. 3: 387–418.

McKeon, Michael. 1987. *The Origins of the English Novel, 1600–1740*. Baltimore, MD: Johns Hopkins University Press.

McMahon, Walter W. 2009. *Higher Learning, Greater Good: The Private and Social Benefits of Higher Education*. Baltimore, MD: Johns Hopkins University Press.

Meekings, Kiran, Cory Williams, and John Arrowsmith. 2012. "Orphan Drug Development: An Economically Viable Strategy for Biopharma R&D." *Drug Discovery Today* 17, no. 13: 660–64.

Megill, Allan. 1994. *Rethinking Objectivity*. Durham, NC: Duke University Press.

Mena, Bryan. 2021. "Gov. Greg Abbott and Other Republicans Blamed Green Energy for Texas' Power Woes: But the State Runs on Fossil Fuels." *Texas Tribune*, February 17. https://www.texastribune.org/2021/02/17/abbott-republicans-green-energy/.

Menand, Louis, Paul Reitter, and Chad Wellmon. 2016. *The Rise of the Research University: A Sourcebook*. Chicago: University of Chicago Press.

Meranze, Michael. 2015. "Humanities out of Joint." *American Historical Review* 120, no. 4: 1311–26.

Merriman, Ben. 2015. "A Science of Literature." *Boston Review*, August 3. http://bostonreview.net/books-ideas/ben-merriman-moretti-jockers-digital-humanities.

Merry, Sally Engle. 2011. "Measuring the World: Indicators, Human Rights, and Global Governance." *Current Anthropology* 52, no. S3: S83–S93.

———. 2016. *The Seductions of Quantification: Measuring Human Rights, Gender Violence, and Sex Trafficking*. Chicago: University of Chicago Press.

Miller, Margaret A. 2012. "From Denial to Acceptance: The Stages of Assessment." National Institute for Learning Outcomes Assessment Occasional Paper No. 13 (January).

Mimno, David. 2013. "jsLDA." GitHub release with MIT License. https://github.com/mimno/jsLDA.

Mirowski, Philip. 2013. *Never Let a Serious Crisis Go to Waste: How Neoliberalism Survived the Financial Meltdown*. London: Verso.

Mitchell, Polly, and Anna Alexandrova. 2020. "Well-Being and Pluralism." *Journal of Happiness Studies*. https://doi.org/10.1007/s10902-020-00323-8.

Mizoguchi, Hirokuni, Takayuki Yamanaka, and Shingo Kano. 2016. "Research and Drug Development Activities in Rare Diseases: Differences between Japan and Europe Regarding Influence of Prevalence." *Drug Discovery Today* 21, no. 10 (2016): 1681–89.

Moi, Toril. 2017. *Revolution of the Ordinary: Literary Studies after Wittgenstein, Austin, and Cavell*. Durham, NC: Duke University Press.

Montgomery, Catherine. 2016. "From Standardization to Adaptation: Clinical Trials and the Moral Economy of Anticipation." *Science as Culture* 26, no. 2: 232–54.

Moretti, Enrico. 2004. "Estimating the Social Return to Higher Education: Evidence from Longitudinal and Repeated Cross-Sectional Data." *Journal of Econometrics* 121, no. 1–2: 175–212.

Moretti, Franco. 2000. "Conjectures on World Literature." *New Left Review* 1 (January–February): 54–68.

———. 2006. "The End of the Beginning." *New Left Review* 41 (September–October): 71–86.

———. 2007. *Graphs, Maps, Trees: Abstract Models for a Literary History*. London: Verso.

———. 2009a. "Critical Response II: 'Relatively Blunt.'" *Critical Inquiry* 36, no. 1: 172–74.

———. 2009b. "Style, Inc. Reflections on Seven Thousand Titles (British Novels, 1740–1850)." *Critical Inquiry* 36, no. 1: 134–58.

———. 2013. *Distant Reading*. London: Verso.

Morgan, Mary S. 2007. "The Curious Case of the Prisoner's Dilemma: Model Situation? Exemplary Narrative?" In *Science without Laws: Model Systems, Cases, Exemplary Narratives*, edited by Angela N. H. Creager, Elizabeth Lunbeck, and M. Norton Wise, 157–88. Durham, NC: Duke University Press.

Morgan, Mary. S., and Maria Bach. 2018. "Measuring Development—from the UN's Perspective." *History of Political Economy* 50, no. S1: 193–210.

Moscati, Ivan. 2013. "How Cardinal Utility Entered Economic Analysis: 1909–1944." *European Journal of the History of Economic Thought* 20, no. 6: 906–39.

Mounk, Yascha. 2018. *The People vs. Democracy: Why Our Freedom Is in Danger and How to Save It*. Cambridge, MA: Harvard University Press.

Mudde, Cas. 2004. "The Populist Zeitgeist." *Government and Opposition* 39, no. 4: 542–63.

Müller, Jan-Werner. 2016. *What Is Populism?* Philadelphia: University of Pennsylvania Press.

Muller, Jerry Z. 2018. *The Tyranny of Metrics*. Princeton, NJ: Princeton University Press.

Nakassis, Constantine V. "Brands and Their Surfeits." *Cultural Anthropology* 28, no. 1 (2013): 111–26.

National Academies of Sciences, Engineering, and Medicine. 2016. *Attribution of Extreme Weather Events in the Context of Climate Change*. Washington, DC: National Academies Press. http://www.nap.edu/catalog/21852.

National Commission on Excellence in Education. 1983. *A Nation at Risk: The Imperative for Educational Reform*. Washington, DC: US Government Printing Office.

National Economic Council and Office of Science and Technology Policy. 2015. "A Strategy for American Innovation." Obama White House Archives. https://obamawhitehouse.archives.gov/sites/default/files/strategy_for_american_innovation_october_2015.pdf.

National Institute for Health and Care Excellence (NICE). 2008. "Social Value Judgements: Principles for the Development of NICE Guidance." Version 2. National Center for Biotechnology Information, July. https://www.ncbi.nlm.nih.gov/books/NBK395865/.

———. 2013. "Guide to the Methods of Technology Appraisal 2013." https://www.nice.org.uk/process/pmg9/chapter/foreword.

———. 2016. "Specification for Cancer Drugs Fund Data Collection Arrangements." Centre for Health Technology Evaluation, Technology Appraisals Programme. https://www.nice.org.uk/Media/Default/About/what-we-do/NICE-guidance/NICE-technology-appraisal-guidance/cancer-drugs-fund/data-collection-specification.pdf.

Natsis, Yannis. 2016. "Scientists Voice Concerns about Adaptive Pathways." European Public Health Alliance, May 26. https://epha.org/scientists-voice-concerns-about-adaptive-pathways/.

Nelson, Laura. 2020. "TAMIDS Seminar Laura Nelson 2020 10 02." Lecture, Texas A&M Institute of Data Science, October 29. YouTube video, 42:54. https://youtu.be/UPjbJztuUoo.

Newfield, Christopher. 1996. *The Emerson Effect: Individualism and Submission in America*. Chicago: University of Chicago Press.

———. 2003. *Ivy and Industry: Business and the Making of the American University, 1880–1980*. Durham, NC: Duke University Press.

———. 2008. *Unmaking the Public University: The Forty Year Assault on the Middle Class*. Cambridge, MA: Harvard University Press.

———. 2016. *The Great Mistake: How We Wrecked Public Universities and How We Can Fix Them*. Baltimore, MD: Johns Hopkins University Press.

Newman, John Henry. 1996. *The Idea of a University*. Edited by Frank Turner. New Haven, CT: Yale University Press.

NHS England Cancer Drugs Fund Team. 2016. *Appraisal and Funding of Cancer Drugs from July 2016 (Including the New Cancer Drugs Fund): A New Deal for Patients, Taxpayers and Industry*. London: NHS England. https://www.england.nhs.uk/wp-content/uploads/2013/04/cdf-sop.pdf.

Nichols, Tom. 2020. "A Large Portion of the Electorate Chose the Sociopath." *The Atlantic*, November 4. https://www.theatlantic.com/ideas/archive/2020/11/large-portion-electorate-chose-sociopath/616994/.

Nirenberg, David, and Ricardo L. Nirenberg. 2021. *Uncountable: A Philosophical History of Number and Humanity from Antiquity to the Present*. Chicago: University of Chicago Press.

Noble, Safiya Umoja. 2018. *Algorithms of Oppression: How Search Engines Reinforce Racism*. New York: New York University Press.

Norris, Pippa. 2011. *Democratic Deficit: Critical Citizens Revisited*. Cambridge: Cambridge University Press.

Nullis, Clare. 2018. "Greenhouse Gas Levels in Atmosphere Reach New Record." World Meteorological Organization, November 20. https://public.wmo.int/en/media/press-release/greenhouse-gas-levels-atmosphere-reach-new-record.

Nussbaum, Martha. 1992. "Human Functioning and Social Justice: In Defense of Aristotelian Essentialism." *Political Theory* 20, no. 2: 202–46.

Nuwer, Rachel. 2012. "Data Mining the Classics Clusters Women Authors Together, Puts Melville out on a Raft." *Smithsonian Magazine*, Smart News, August 27. http://www.smithsonianmag.com/smart-news/data-mining-the-classics-clusters-women-authors-together-puts-mellville-out-on-a-raft-16028354/.

O'Carroll, Eoin. 2018. "Can Your Boss Make You Wear a Fitness Tracker?" *Christian Science Monitor*, March 15. https://www.csmonitor.com/Technology/2018/0315/Can-your-boss-make-you-wear-a-Fitbit.

O'Connor, Brendon. 2002. "Policies, Principles, and Polls: Bill Clinton's Third Way Welfare Politics, 1992–1996." *Australian Journal of Politics and History* 48, no. 3: 396–411.

Office for National Statistics (ONS). 2012. "Proposed Domains and Headline Indicators for Measuring National Well-Being." http://www.ons.gov.uk/ons/about-ons/get-involved/consultations/archived-consultations/2012/measuring-national-well-being-domains/index.html.

———. 2018. "GDP First Quarterly Estimate, UK: April to June 2018." https://www.ons.gov.uk/economy/grossdomesticproductgdp/bulletins/gdpfirstquarterly estimateuk/apriltojune2018#toc.

———. 2019. "Measures of National Well-Being Dashboard, October 23, 2019." https://www.ons.gov.uk/peoplepopulationandcommunity/wellbeing/articles/measuresofnationalwellbeingdashboard/2018-04-25.

Ogden, Thomas. 1997. *Reverie and Interpretation: Sensing Something Human.* New York: Jason Aronson.

Oishi, Shigehiro, Ulrich Schimmack, and Stanley J. Colcombe. 2003. "The Contextual and Systematic Nature of Life Satisfaction Judgments." *Journal of Experimental Social Psychology* 39, no. 3: 232–47.

Olsen, Henry. 2021. "New Poll: There Is No Singular Trump Voter." *Washington Post,* February 5. https://www.washingtonpost.com/opinions/2021/02/05/new-poll -there-is-no-singular-trump-voter/.

O'Neil, Cathy. 2016. *Weapons of Math Destruction.* New York: Crown.

O'Neill, Onora, Jocelyn Cornwell, Alastair Thompson, and Charles Vincent. 2008. *Safer Births: Everybody's Business.* London: King's Fund.

Oreopoulos, Philip, and Kjell G. Salvanes. 2011. "Priceless: The Nonpecuniary Benefits of Schooling." *Journal of Economic Perspectives* 25: 159–84.

Oreskes, Naomi. 2004. "The Scientific Consensus on Climate Change." *Science* 306, no. 5702: 1686–86.

Organisation for Economic Co-operation and Development (OECD). 2013. *OECD Guidelines on Measuring Subjective Well-Being.* Paris: OECD iLibrary. http://dx .doi.org/10.1787/9789264191655-en.

———. 2017. *How's Life? 2017: Measuring Well-Being.* Paris: OECD iLibrary. https:// doi.org/10.1787/how_life-2017-en.

O'Rourke, Brian, Wija Oortwijn, and Tara Schuller. 2020. "Announcing the New Definition of Health Technology Assessment." *Value Health* 23, no. 6: 824–25. https://www.valueinhealthjournal.com/article/S1098-3015(20)32060-X/pdf.

Otto, Friederike E. L., Geert J. van Oldenborgh, Jonathan Eden, David J. Karoly, Peter A. Stott, and Myles R. Allen. 2016. "Framing the Question of Attribution of Extreme Weather Events." *Nature Climate Change* 6, no. 9: 813–16.

Parker, Wendy S. 2009. "Confirmation and Adequacy-for-Purpose in Climate Modelling." *Aristotelian Society Supplementary* 83, no. 1: 233–49. https://doi.org /10.1111/j.1467-8349.2009.00180.x.

———. 2010. "Whose Probabilities? Predicting Climate Change with Ensembles of Models." *Philosophy of Science* 77, no. 5: 985–97. https://doi.org/10.1086 /656815.

———. 2014. "Values and Uncertainties in Climate Prediction, Revisited." *Studies in History and Philosophy of Science Part A* 46: 24–30.

Pearson, Steven. 2007. "Standards of Evidence." In *The Learning Healthcare System: Workshop Summary of the Roundtable on Evidence-Based Medicine,* edited by LeighAnne Olsen, Dara Aisner, and J. Michael McGinnis, 171–74. Washington DC: National Academies Press.

Perry, Tim. 2016. "Clinton Sets the Record Straight on Her Free Trade Stance." CBS News, May 9. https://www.cbsnews.com/news/clinton-sets-the-record-straight -on-her-free-trade-stance/.

Peters, Chris. 2014. "5 a Day All over the World?" Sense about Science, January 29. http://archive.senseaboutscience.org/blog.php/81/5-a-day-all-over-the-world .html.

Philipsen, Dirk. 2015. *The Little Big Number: How GDP Came to Rule the World and What to Do about It.* Princeton, NJ: Princeton University Press.

Phillips, Adam. 1994. "The Experts." *London Review of Books* 16, no. 24. https://www .lrb.co.uk/the-paper/v16/n24/adam-phillips/the-experts.

Phillips, Chuck. 1992. "'I Do Not Advocate . . . Murdering': 'Raptivist' Sister Souljah Disputes Clinton Charge." *Los Angeles Times*, June 17. https://www.latimes.com/archives/la-xpm-1992-06-17-ca-573-story.html.

Piketty, Thomas. 2014. *Capital in the Twenty-First Century*. Cambridge, MA: Harvard University Press.

Piper, Andrew. 2018. *Enumerations: Data and Literary Study*. Chicago: University of Chicago Press.

———. 2020a. *Can We Be Wrong? The Problem of Textual Evidence in a Time of Data*. New York: Cambridge University Press.

———. 2020b. "Do We Know What We're Doing?" *Journal of Cultural Analytics* 1, no. 1. https://doi.org/10.22148/001c.11826.

Pogge, Thomas. 2002. "Can the Capability Approach Be Justified?" *Philosophical Topics* 30, no. 2: 167–228.

Poniewozik, James. 2019. *Audience of One: Donald Trump, Television, and the Fracturing of America*. New York: Liveright.

Poovey, Mary. 1998. *A History of the Modern Fact: Problems of Knowledge in the Sciences of Wealth and Society*. Chicago: University of Chicago Press.

Porter, Theodore M. 1992. "Quantification and the Accounting Ideal in Science." *Social Studies of Science* 22, no. 4: 633–51.

———. 1994. "Making Things Quantitative." *Science in Context* 7, no. 3: 389–407.

———. 1995. *Trust in Numbers: The Pursuit of Objectivity in Science and Public Life*. Princeton, NJ: Princeton University Press.

Power, Michael. 1997. *The Audit Society: Rituals of Verification*. Oxford: Oxford University Press.

Premier Research. n.d. "Mapping the New Landscape of Orphan Drug Development." Accessed February 26, 2021. https://premier-research.com/mapping-the-new-landscape-of-orphan-drug-development/.

Prendergast, Christopher. 2005. "Evolution and Literary History: A Response to Franco Moretti." *New Left Review* 34 (July–August): 40–62.

President's Commission on Higher Education. 1947. *Higher Education for Democracy: A Report of the President's Commission on Higher Education*. Vol. 1. New York: Harper and Brothers.

Pritchett, Lant. 2001. "Where Has All the Education Gone?" *World Bank Economic Review* 15, no. 3: 367–91.

———. 2006. "Does Learning to Add Up Add Up? The Returns to Schooling in Aggregate Data" In *Handbook of the Economics of Education*, vol. 1, edited by Eric Hanushek and Finis Welch, 635–95. Amsterdam: North-Holland.

Ragin, Charles C., and Howard S. Becker. 1992. "Introduction: Cases of 'What Is a Case?'" In *What Is a Case? Exploring the Foundations of Social Inquiry*, edited by Charles C. Ragin and Howard S. Becker, 3–18. Cambridge: Cambridge University Press.

Ramsay, Steve. 2011. *Reading Machines: Toward an Algorithmic Criticism*. Champaign: University of Illinois Press.

Rasi, Guido, and Hans-Georg Eichler. 2016. Letter to Silvio Garattini et al., June 16. European Medicines Agency. http://www.ema.europa.eu/docs/en_GB/document_library/Other/2016/06/WC500208968.pdf.

Rawlins, Michael, David Barnett, and Andrew Stevens. 2010. "Pharmacoeconomics: NICE's Approach to Decision-Making." *British Journal of Clinical Pharmacology* 70, no. 3: 346–49.

Rawls, John. 1993. *Political Liberalism*. New York: Columbia University Press.

———. 1999. *A Theory of Justice*. Rev. ed. Cambridge, MA: Harvard University Press.

Ray, Rashawn, and William A. Galston. 2020. "Did the 1994 Crime Bill Cause Mass Incarceration?" Brookings, August 28. https://www.brookings.edu/blog/fixgov /2020/08/28/did-the-1994-crime-bill-cause-mass-incarceration/.

Rees, Clea. 2014. "Better Lie!" *Analysis* 74, no. 1: 69–74.

Reeves, Richard V. 2018. *Dream Hoarders: How the American Upper Middle Class Is Leaving Everyone Else in the Dust, Why That Is a Problem, and What to Do about It*. Washington, DC: Brookings Institution Press.

Reich, Robert. 1991. *The Work of Nations*. New York: Vintage.

Reiss, Julian. 2016. *Error in Economics: Towards a More Evidence-Based Methodology*. London: Routledge.

Reynolds, Glenn. 2017. "The Suicide of Expertise." *USA Today*, March 20.

Richardson, Henry. 1990. "Specifying Norms as a Way to Resolve Concrete Ethical Problems." *Philosophy and Public Affairs* 19, no. 4: 279–310.

Robinson, Cedric J. 2007. *Forgeries of Memory and Meaning: Blacks and the Regimes of Race in American Theater and Film before World War II*. Chapel Hill: University of North Carolina Press.

———. 2016. *Terms of Order*. Chapel Hill: University of North Carolina Press.

Rockwell, Geoffrey, and Stéfan Sinclair. 2016. *Hermeneutica: Computer-Assisted Interpretation in the Humanities*. Cambridge, MA: MIT Press.

Rodrik, Dani. 2006. "Goodbye Washington Consensus, Hello Washington Confusion? A Review of the World Bank's *Economic Growth in the 1990s: Learning from a Decade of Reform*." *Journal of Economic Literature* 44, no. 4: 973–87.

Rogers, Wendy A., and Mary Jean Walker. 2017. "The Line-Drawing Problem in Disease Definition." *Journal of Medicine and Philosophy* 42, no. 4: 405–23.

Romer, Paul M. 1990. "Endogenous Technological Change." *Journal of Political Economy* 98, no. 5: S71–S102.

Rorty, Richard. 1989. *Contingency, Irony and Solidarity*. Cambridge: Cambridge University Press.

Rosenthal, Laura J. 2011. "Assessment, Literary Study, and Disciplinary Futures." In *Literary Study, Measurement, and the Sublime: Disciplinary Assessment*, edited by Donna Heiland and Laura J. Rosenthal, 183–98. New York: Teagle Foundation.

Ross, Shawna. 2014. "In Praise of Overstating the Case: A Review of Franco Moretti, *Distant Reading* (Verso 2013)." *Digital Humanities Quarterly* 8, no. 1. http://www .digitalhumanities.org/dhq/vol/8/1/000171/000171.html.

Rottenburg, Richard, Sally E. Merry, Sun-Joon Park, and Johanna Mugler, eds. 2015. *The World of Indicators: The Making of Governmental Knowledge through Quantification*. Cambridge: Cambridge University Press.

Rottenburg, Richard, and Sally Engle Merry. 2015. "A World of Indicators: The Making of Governmental Knowledge through Indicators." In *The World of Indicators: The Making of Governmental Knowledge through Quantification*, edited by Richard Rottenburg, Sally E. Merry, Sung-Joon Park, and Johanna Mugler, 1–33. Cambridge: Cambridge University Press.

Rumberger, Russell W. 1987. "The Impact of Surplus Schooling on Productivity and Earnings." *Journal of Human Resources* 22, no. 1: 24–50.

Rustin, Susanna. 2012. "Can Happiness Be Measured?" Interview with Richard Layard and Julian Baggini. *The Guardian*, July 20. https://www.theguardian.com /commentisfree/2012/jul/20/wellbeing-index-happiness-julian-baggini.

Rybicki, Jan. 2016. "*Vive la différence*: Tracing the (Authorial) Gender Signal by Multivariate Analysis of Word Frequencies." *Digital Scholarship in the Humanities* 31, no. 4: 746–61.

Sala-i-Martin, Xavier X. 1997. "I Just Ran Four Million Regressions." National Bureau of Economic Research Working Paper 6252. https://www.nber.org/papers/w6252.

Salvanto, Anthony, Jennifer de Pinto, Fred Backus, Kabir Khanna, and Elena Cox. 2020. "CBS News Poll: Most Feel Election Is 'Settled' but Trump Voters Disagree." CBS News, December 13. https://www.cbsnews.com/news/cbs-news-poll-most -feel-election-is-settled-but-trump-voters-disagree/.

Samuels, David. 2016. "The Aspiring Novelist Who Became Obama's Foreign-Policy Guru." *New York Times Magazine*, May 5. https://www.nytimes.com/2016/05/08 /magazine/the-aspiring-novelist-who-became-obamas-foreign-policy-guru.html.

Samuelson, Paul A., and William D. Nordhaus. 2005. *Economics*. 18th ed. New York: McGraw-Hill Irwin.

Sass, Louis A. 2004. "'Negative Symptoms,' Common Sense and Cultural Disembedding in the Modern Age." In *Schizophrenia, Culture, and Subjectivity: The Edge of Experience*, edited by Janis Hunter Jenkins and Robert Barrett, 303–28. Cambridge: Cambridge University Press.

Sasson, Eric. 2016. "Blame Trump's Victory on College-Educated Whites, Not the Working Class." *New Republic*, November 15. https://newrepublic.com/ article/138754/blame-trumps-victory-college-educated-whites-not-working-class.

Saul, Jennifer. 2012. *Lying, Misleading, and What Is Said*. Oxford: Oxford University Press.

Schmidt, Alexander. 2013. "Self-Cultivation (Bildung) and Sociability between Mankind and the Nation: Fichte and Schleiermacher on Higher Education." In *Ideas of Education: Philosophy and Politics from Plato to Dewey*, edited by Christopher Brooke and Elizabeth Frazer, 160–77. London: Routledge.

Schmidt, Ben. 2019. "A Computational Critique of a Computational Critique of Computational Critique." March 18. http://benschmidt.org/post/critical_inquiry /2019-03-18-nan-da-critical-inquiry/.

Schroeder, Carsten, and Shlomo Yitzhaki. 2017. "Revisiting the Evidence for Cardinal Treatment of Ordinal Variables." *European Economic Review* 92: 337–58.

Schwaber, Evelyne. 1983. "Psychoanalytic Listening and Psychoanalytic Reality." *International Review of Psycho-Analysis* 10: 379–92.

Schwarz, Norbert, and Fritz Strack. 1991. "Evaluating One's Life: A Judgment Model of Subjective Well-Being." In *Subjective Well-Being: An Interdisciplinary Perspective*, edited by Fritz Strack, Michael Argyle, and Norbert Schwarz, 27–47. Elmsford, NY: Pergamon.

———. 1999. "Reports of Subjective Well-Being: Judgmental Processes and Their Methodological Implications." In *Well-Being: The Foundations of Hedonic Psychology*, edited by Daniel Kahnemann, Ed Diener, and Norbert Schwarz, 61–84. New York: Russell Sage Foundation.

Sculley, D., and Bradley M. Pasanek. 2008. "Meaning and Mining: The Impact of Implicit Assumptions in Data Mining for the Humanities." *Literary and Linguistic Computing* (now *Digital Studies in Research*) 23, no. 4: 409–24.

Seife, Charles. 2010. *Proofiness: How You're Being Fooled by the Numbers*. New York: Viking.

Seligman, Martin E. P. 2004. *Authentic Happiness: Using the New Positive Psychology to Realize Your Potential for Lasting Fulfillment*. New York: Simon and Schuster.

Seligman, Martin E. P., and Mihaly Csikszentmihalyi. 2000. "Happiness, Excellence, and Optimal Human Functioning." Special issue, *American Psychologist* 55, no. 1: 5–183.

Sen, Amartya. 1987. *Ethics and Economics*. Oxford: Blackwell.

———. 2004. "Elements of a Theory of Human Rights." *Philosophy and Public Affairs* 32, no. 4: 315–56.

Shapin, Steven, and Simon Schaffer. 1985. *Leviathan and the Air-Pump*. Princeton, NJ: Princeton University Press.

Shapiro, Jesse M. 2006. "Smart Cities: Quality of Life, Productivity, and the Growth Effects of Human Capital." *Review of Economics and Statistics* 88: 324–35.

Shore, Cris, and Susan Wright. 2015. "Audit Culture Revisited: Rankings, Ratings, and the Reassembling of Society." *Current Anthropology* 56, no. 3: 421–44.

Silver, Nate. 2016. "Education, Not Income, Predicted Who Would Vote for Trump." FiveThirtyEight, November 22. http://fivethirtyeight.com/features/education -not-income-predicted-who-would-vote-for-trump/.

Simonite, Tom. 2021. "The Exile." *Wired* (July/August): 114–27.

Simpson, Ian, Kelsey Beninger, and Rachel Ormston. 2015. "Public Confidence in Official Statistics." NatCen UK, February 23. https://natcen.ac.uk/media/833802 /public-confidence-in-official-statistics_-final.pdf.

Sinclair, Stéfan, and Geoffrey Rockwell. n.d. Voyant Tools. Accessed August 5, 2021. http://www.voyant-tools.org.

Singh, Ramnadeep, and Anna Alexandrova. 2020. "Happiness Economics as Technocracy." *Behavioural Public Policy* 4, no. 22: 236–44.

Skowronek, Stephen. 1997. *The Politics Presidents Make: Leadership from John Adams to Bill Clinton*. Cambridge, MA: Belknap Press of Harvard University Press.

Smith, Adam. (1759) 1853. *The Theory of Moral Sentiments; or, An Essay towards an Analysis of the Principles by which Men naturally judge concerning the Conduct and Character, first of their Neighbours, and afterwards of themselves. To which is added, A Dissertation on the Origins of Languages*. Edited by Dugald Stewart. New Edition. London: Henry G. Bohn.

Smith, Erika D. 2015. "Why I'm Leaving Indianapolis." *Indianapolis Star*, April 24. https://www.indystar.com/story/opinion/columnists/erika-smith/2015/04/23 /smith-leaving-indianapolis/26233799/.

Smyth, Chris 2017. "Three Portions of Raw Fruit-and-Veg as Healthy as Five a Day." *The Times*, August 30. https://www.thetimes.co.uk/article/three-portions-of-raw -fruit-and-veg-as-healthy-as-five-a-day-cnzhncdqb.

Snow, Charles Percy. (1959) 2001. *The Two Cultures*. Cambridge: Cambridge University Press.

Solow, Robert M. 1957. "Technical Change and the Aggregate Production Function." *Review of Economics and Statistics* 39, no. 3: 312–20.

Spence, Michael. 1973. "Job Market Signaling." *Quarterly Journal of Economics* 87, no. 3: 355–74.

Stableford, Dylan. 2016. "Donald Trump: 'I Love the Poorly Educated.'" Yahoo News, February 24. http://news.yahoo.com/trump-i-love-the-poorly-educated -144008662.html.

Steel, Daniel. 2016. "Climate Change and Second-Order Uncertainty: Defending a Generalized, Normative, and Structural Argument from Inductive Risk." *Perspectives on Science* 24, no. 6: 696–21. https://doi.org/10.1162/POSC_a_00229.

Stevenson, Betsy, and Jonathan Wolfers. 2008. "Economic Growth and Subjective Well-Being: Reassessing the Easterlin Paradox." National Bureau of Economic Research Working Paper w14282. https://papers.ssrn.com/sol3/papers.cfm?abstract_id=1261469.

Stiglitz, Joseph E., Amartya Sen, and Jean-Paul Fitoussi. 2009. *Report by the Commission on the Measurement of Economic Performance and Social Progress.* http://citeseerx.ist.psu.edu/viewdoc/download?doi=10.1.1.215.58&rep=rep1&type=pdf.

Stocker, Thomas F., Dahe Qin, Gian-Kasper Plattner, Melinda M. B. Tignore, Simon K. Allen, Judith Boschung, Alexander Nauels, Yu Xia, Vincent Bex, and Pauline M. Midgley. 2013. "Summary for Policymakers, Technical Summary and Frequently Asked Questions." In *Climate Change 2013: The Physical Science Basis.* Contribution of Working Group I to the Fifth Assessment Report of the Intergovernmental Panel on Climate Change. Geneva: Intergovernmental Panel on Climate Change.

Stoker, Gerry. 2006. *Why Politics Matters: Making Democracy Work.* Basingstoke, UK: Palgrave Macmillan.

Stone, Arthur A., Stefan Schneider, Alan Krueger, Joseph E. Schwartz, and Angus Deaton. 2016. "Experiential Wellbeing Data from the American Time Use Survey: Comparisons with Other Methods and Analytic Illustrations with Age and Income." *Social Indicators Research* 136, no. 1: 1–20.

Stone, Dáithí A., and Myles R. Allen. 2005. "The End-to-End Attribution Problem: From Emissions to Impacts." *Climatic Change* 71, no. 3: 303–18.

Stott, Peter A., Myles Allen, Nikolaos Christidis, Randall M. Dole, Martin Hoerling, Chris Huntingford, Pardeep Pall, Judith Perlwitz, and Dáithí Stone. 2013. "Attribution of Weather and Climate-Related Events." In *Climate Science for Serving Society: Research, Modeling and Prediction Priorities*, edited by Ghassem R. Asrar and James W. Hurrell, 307–37. Dordrecht: Springer. http://link.springer.com/chapter/10.1007/978-94-007-6692-1_12.

Strathern, Marilyn, ed. 2000. *Audit Cultures: Anthropological Studies in Accountability, Ethics, and the Academy.* London: Routledge.

Streeck, Wolfgang. 2014. *Buying Time: The Delayed Crisis of Democratic Capitalism.* London: Verso.

Sunstein, Cass. 2014. *Why Nudge? The Politics of Libertarian Paternalism.* New Haven, CT: Yale University Press.

Suspitsyna, Tatiana. 2010. "Accountability in American Education as a Rhetoric and Technology of Governmentality." *Journal of Education Policy* 25, no. 5: 567–86.

Swan, Elaine. 2010. "States of White Ignorance, and Audit Masculinity in English Higher Education." *Social Politics: International Studies in Gender, State and Society* 17, no. 4: 477–506.

Tal, Eran. 2013. "Old and New Problems in Philosophy of Measurement." *Philosophy Compass* 8, no. 12: 1159–73.

Tambuyzer, Erik. 2010. "Rare Diseases, Orphan Drugs and Their Regulation: Questions and Misconceptions." *Nature Reviews Drug Discovery* 9, no. 12: 921–29.

Taussig, Karen-Sue, Klaus Hoeyer, and Stefan Helmreich. 2013. "The Anthropology of Potentiality in Biomedicine: An Introduction to Supplement 7." *Current Anthropology* 54, no. S7: S3–S14.

Telegraph Reporters. 2017. "Eat 10 Fruit-and-Veg a Day for a Longer Life, Not Five." *Daily Telegraph*, February 23.

Thompson, Allen, and Friederike E. L. Otto. 2015. "Ethical and Normative Implications of Weather Event Attribution for Policy Discussions Concerning Loss and Damage." *Climatic Change* 133, no. 3: 439–51. https://doi.org/10.1007/s10584-015-1433-z.

Thomson Reuters. 2012. "The Economic Power of Orphan Drugs." http://dtr-pharma .com/wp-content/uploads/2018/09/THE-ECONOMIC-POWER-of-Orphan -Drugs.pdf.

Timmermans, Stefan, and Marc Berg. 2003. *The Gold Standard: An Exploration of Evidence-Based Medicine and Standardization in Health Care*. Philadelphia: Temple University Press.

Tooze, Adam. 2018. *Crashed: How a Decade of Financial Crises Changed the World*. London: Penguin.

Toscano, Alberto. 2017. "Notes on Late Fascism." *Historical Materialism*, April 2. http://www.historicalmaterialism.org/blog/notes-late-fascism#_ftn2.

Trenberth, Kevin E., John T. Fasullo, and Theodore G. Shepherd. 2015. "Attribution of Climate Extreme Events." *Nature Climate Change* 5, no. 8: 725–30.

Treurniet, Nikolaas. 1997. "On an Ethic of Psychoanalytic Technique." *Psychoanalytic Quarterly* 66, no. 4: 596–627.

Trumpener, Katie. 2009. "Critical Response I: Paratext and Genre System: A Response to Franco Moretti." *Critical Inquiry* 36, no. 1: 159–71.

Tuana, Nancy. 2013. "Embedding Philosophers in the Practices of Science: Bringing Humanities to the Sciences." *Synthese* 190, no. 11: 1955–73.

Tuana, Nancy, Ryan L. Sriver, Toby Svoboda, Roman Olson, Peter J. Irvine, Jacob Haqq-Misra, and Klaus Keller. 2012. "Towards Integrated Ethical and Scientific Analysis of Geoengineering: A Research Agenda." *Ethics, Policy and Environment* 15, no. 2: 136–57.

Tung, Charles M. 2011. "The Future of Literary Criticism: Assessment, the Curricularized Classroom, and Thick Reading." In *Literary Study, Measurement, and the Sublime: Disciplinary Assessment*, edited by Donna Heiland and Laura J. Rosenthal, 199–216. New York: Teagle Foundation.

Underwood, Ted. 2013. "We Don't Already Understand the Broad Outlines of Literary History." *The Stone and the Shell* (blog), February 8. https://tedunderwood .com/2013/02/08/we-dont-already-know-the-broad-outlines-of-literary-history/.

———. 2017. "A Genealogy of Distant Reading." *Digital Humanities Quarterly* 11, no. 2. http://www.digitalhumanities.org/dhq/vol/11/2/000317/000317.html.

———. 2018. Talk on the panel "DH Projects in Teaching and Research," at "Digital Humanities Nuts and Bolts: From Idea to Sustainable Project." National Humanities Center, Research Triangle Park, NC, October 2.

———. 2019. *Distant Horizons: Digital Evidence and Literary Change*. Chicago: University of Chicago Press.

United Nations Development Programme (UNDP). n.d. "Goal 10: Reduced Inequalities." Accessed June 5, 2019. https://www.undp.org/content/undp/en/home /sustainable-development-goals/goal-10-reduced-inequalities.html.

Urciuoli, Bonnie. 2005. "The Language of Higher Education Assessment: Legislative Concerns in a Global Context." *Indiana Journal of Global Legal Studies* 12, no. 1: 183–204.

US Department of Education. 2006. *A Test of Leadership: Charting the Future of US Higher Education*. Washington, DC: US Department of Education.

US Departments of Justice and Education. 2011. *Guidance on the Voluntary Use of Race to Achieve Diversity and Avoid Racial Isolation in Elementary and Secondary Schools*. Washington, DC: US Department of Education.

US Food and Drug Administration (FDA). n.d. Search Orphan Drug Designations and Approvals. Data retrieved February 3, 2021. https://www.accessdata.fda.gov/scripts/opdlisting/oopd.

US National Institute of Health (NIH). 2021. "FAQ about Rare Diseases." Genetic and Rare Diseases Information Center, January 26 version. https://rarediseases.info.nih.gov/diseases/pages/31/faqs-about-rare-diseases.

Valero, Anna, and John Van Reenen. 2019. "The Economic Impact of Universities: Evidence from across the Globe." *Economics of Education Review* 68: 53–67.

Vallier, Kevin. 2018. "Public Justification." In *Stanford Encyclopedia of Philosophy*, edited by Edward N. Zalta. Spring edition. https://plato.stanford.edu/archives/spr2018/entries/justification-public.

van Fraassen, Bas C. 2008. *Scientific Representation: Paradoxes of Perspective*. Oxford: Oxford University Press.

Vermeule, Blakey. 2009. *Why Do We Care about Literary Characters?* Baltimore, MD: Johns Hopkins University Press.

———. 2012. "Critical Response VI: Wit and Poetry and Pope, or The Handicap Principle." *Critical Inquiry* 38, no. 2: 426–30.

Vessonen, E. 2019. "Representing and Constructing: Psychometrics from the Perspectives of Measurement Theory and Concept Formation." PhD diss., University of Cambridge.

Waldron, Jeremy. 1987. "Theoretical Foundations of Liberalism." *The Philosophical Quarterly* 37, no. 147: 127–50.

Webber, Jonathan. 2013. "Liar!" *Analysis* 73, no. 4: 651–59.

Webster, Andrew. 2012. "Introduction: Bio-Objects: Exploring the Boundaries of Life." In *Bio-Objects*, edited by Niki Vermeulen, Sakari Tamminen, and Andrew Webster, 15–24. Abingdon, UK: Routledge.

———. 2013. "Introduction: The Boundaries and Mobilities of Regenerative Medicine." In *The Global Dynamics of Regenerative Medicine*, edited by Andrew Webster, 1–17. London: Palgrave Macmillan.

Wedeen, Lisa. 1998. "Acting 'As If': Symbolic Politics and Social Control in Syria." *Comparative Studies in Society and History* 40, no. 3: 503–23.

Weed, Elizabeth, and Ellen Rooney, eds. 2014. "In the Shadow of the Digital Humanities." Special issue, *differences* 25, no. 1.

Wellmon, Chad. 2015. *Organizing Enlightenment: Information Overload and the Invention of the Modern Research University*. Baltimore, MD: Johns Hopkins University Press.

Williams, Joan. 2017. "Joan Williams on Misconceptions about the White Working Class." Interview by Michael Krasny. KQED Forum, June 12. https://www.kqed.org/forum/2010101860513/rebroadcast-joan-williams-on-misconceptions-about-the-white-working-class.

Williams, Raymond. 1977. *Marxism and Literature*. Oxford: Oxford University Press.

Williamson, Timothy. 1994. *Vagueness*. Abingdon, UK: Routledge.

Wilsdon, James, Liz Allen, Eleonora Belfiore, et al. 2015. *The Metric Tide: Report of the Independent Review of the Role of Metrics in Research Assessment and Management.* Bristol, UK: Higher Education Funding Council for England.

Winsberg, Eric. 2012. "Values and Uncertainties in the Predictions of Global Climate Models." *Kennedy Institute of Ethics Journal* 22, no. 2: 111–37.

Winsberg, Eric, and William Mark Goodwin. 2016. "The Adventures of Climate Science in the Sweet Land of Idle Arguments." *Studies in History and Philosophy of Science Part B: Studies in History and Philosophy of Modern Physics* 54: 9–17.

Witmer, David R. 1970. "Economic Benefits of College Education." *Review of Educational Research* 40, no. 4: 511–23.

Witte, Griff. 2016. "9 out of 10 Experts Agree: Britain Doesn't Trust the Experts on Brexit." *Washington Post*, June 21. https://www.washingtonpost.com/world /europe/9-out-of-10-experts-agree-britain-doesnt-trust-the-experts-on-brexit /2016/06/21/2ccc134a-34a6-11e6-ab9d-1da2b0f24f93_story.html.

Wollstonecraft, Mary. (1788) 1976. *"Mary" and "The Wrongs of Woman."* Edited by Gary Kelly. New York: Oxford University Press.

Wood, Michael. 2015. "Nothing but the Worst." *London Review of Books* 37, no. 1. https://www.lrb.co.uk/the-paper/v37/n01/michael-wood/nothing-but-the-worst.

Woodhall, Maureen, and Vernon Ward. 1972. *Economic Aspects of Education: A Review of Research in Britain.* Slough, UK: National Foundation for Educational Research.

Wooldridge, Jeffrey M. 2002. *Econometric Analysis of Cross Section and Panel Data.* Cambridge, MA: MIT Press.

World Health Organization (WHO). 2003. *Diet, Nutrition and the Prevention of Chronic Diseases.* Report of a Joint FAO/WHO Expert Consultation. WHO Technical Report Series, No. 916. Geneva: World Health Organization. http://apps .who.int/iris/bitstream/handle/10665/42665/WHO_TRS_916.pdf;jsessionid=B1 BB034463BBAC3A2717F6B52845055E?sequence=1.

———. n.d. "Mental Health." https://www.who.int/news-room/facts-in-pictures/detail /mental-health.

Wring, Dominic. 2005. *The Politics of Marketing the Labour Party.* Basingstoke, UK: Palgrave Macmillan.

Yeh, Rihan. 2016. "Commensuration in a Mexican Border City: Currencies, Consumer Goods, and Languages." *Anthropological Quarterly* 89, no. 1: 63–91.

YouGov. 2017a. "The Economist/YouGov Poll: March 19–21, 2017." https://bit.ly /2PkKjBt.

———. 2017b. "YouGov Survey Results: February 14–15, 2017." https://bit.ly/2M74D7D.

Contributors

ANNA ALEXANDROVA is a Professor in Philosophy of Science at the University of Cambridge. She researches how scientists navigate morally charged and complex phenomena and the role of formal tools, such as models and indicators, in their scholarly and public work. Her articles have appeared in such publications as the *British Journal for the Philosophy of Science*, the *Journal of Moral Philosophy*, and *Behavioural Public Policy*. She is the author of *A Philosophy for the Science of Well-Being* (2017).

GABRIELE BADANO is a Lecturer in the Department of Politics at the University of York. His research interests fall primarily in political philosophy, where he focuses on theories of public reason and liberal democratic self-defense. His articles have appeared in such publications as the *Journal of Political Philosophy*, *Political Studies*, and *Social Theory and Practice*.

ELIZABETH CHATTERJEE is Assistant Professor of Environmental History at the University of Chicago. Her research interests lie at the intersection of energy history, political ecology, and the pathologies of (environmental) policy. Her articles have appeared in such publications as the *Journal of Asian Studies*, *World Development*, and *Development and Change*.

STEPHEN JOHN is the Hatton Trust Senior Lecturer in the Philosophy of Public Health in the Department of History and Philosophy of Science, University of Cambridge, and a fellow of Pembroke College. His recent work is on the ethics and epistemology of chance, categorization, and causation, with a particular focus on the early detection of cancer. His articles have appeared in such publications as *Synthese*, *Philosophical Quarterly*, and *Public Health Ethics*.

TRENHOLME JUNGHANS is a sociocultural anthropologist and Visiting Researcher at the University of Edinburgh. Her research interests include critical analysis of the uses of quantitative and qualitative modalities of knowing, representing, and communicating, with a particular focus on medicine, health care, and the regulation of pharmaceutical products. She received a PhD from the University of St. Andrews, and her articles have appeared in such publications as *Healthcare Analysis* and *Critique of Anthropology.*

GREG LUSK is Assistant Professor of History, Philosophy, and Sociology of Science in the Department of Philosophy and Lyman Briggs College at Michigan State University. His research crosses boundaries between philosophy and climate science by analyzing how climate information can be responsibly produced and used. His articles have appeared in such publications as *Philosophy of Science*, the *British Journal for the Philosophy of Science*, the *Bulletin of the American Meteorological Society*, and *Climatic Change.*

LAURA MANDELL is Director of the Center of Digital Humanities Research and Professor of English at Texas A&M University, where she also directs the Advanced Research Consortium. She is the author of *Breaking the Book: Print Humanities in the Digital Age* (2015) and *Misogynous Economies: The Business of Literature in Eighteenth-Century Britain* (1999), as well as numerous articles, primarily about eighteenth-century women writers.

AASHISH MEHTA received his PhD in Agricultural and Applied Economics from the University of Wisconsin–Madison and is Associate Professor of Global Studies at the University of California, Santa Barbara (UCSB). Prior to joining the UCSB faculty, he worked at the Asian Development Bank. His primary research interest involves the connections between globalization, economic transformation, employment, education, and social stratification. His articles have appeared in such publications as the *International Review of Economics and Finance, Economic Letters*, and the *Cambridge Journal of Economics.*

CHRISTOPHER NEWFIELD is Director of Research at the Independent Social Research Foundation and previously was Distinguished Professor of English at the University of California, Santa Barbara. His interests include US literature before the Civil War and after World War II, critical university studies and critical theory, quantification studies, and the intellectual and social effects of the humanities. His recent book (part three of a trilogy) is *The Great Mistake: How We Wrecked Public Universities and How We Can Fix Them* (2016). He also blogs on higher education policy at Remaking the University.

RAMANDEEP SINGH is a PhD student in the Department of Land Economy at the University of Cambridge and a member of King's College and Trinity Hall. He formerly studied economics, also at Cambridge, and has conducted research for the Brookings Institution. His current research and publications focus on housing and spatial aspects of well-being and the corresponding policy implications, and his articles have appeared in such publications as the *Times Literary Supplement* and *Behavioral Public Policy*.

HEATHER STEFFEN is an affiliated scholar of the Rutgers University Center for Cultural Analysis. She has taught writing at the University of California, Santa Barbara (UCSB), and she has researched metrics and higher education as a postdoctoral scholar at UCSB's Chicano Studies Institute. Her articles on faculty, graduate employee, and undergraduate labor have appeared in such publications as *New Literary History*, *Radical Teacher*, *Cultural Logic*, *Academe*, and the *Chronicle of Higher Education*.

Index

abortion, 49

Academically Adrift (Arum and Roksa), 77

accountability: audits for, 5, 74, 78; clearance rate as metric of, 35; cultures of, 73; learning outcomes assessment for, 71, 79, 85; narratives of advocates of, 72; numerical information associated with, 5, 10, 25, 93, 99–100; as performative, 75; qualitative understanding and, 102; quantocrats lay beyond reach of, 26, 39; relations of, 73, 77, 81, 85; Spellings Report on, 79, 80; unaccountability of "administrative state," 28

accountancy: accounting ideal, 203; benefits of courses in, 238; double-entry, 45n4; Power on meta-accounting, 76, 77, 90; stock market compared with, 41

Adapt Smart initiative, 111

"adequacy for purpose" account, 217n2

affirmative action, 49, 53, 66n2

agency: audit narratives and, 73, 78; *Bildung* and full social, 234; democratic, 91; in knowledge democracy, 64–65; manipulation undermines, 156; quantification erodes, 57, 60; turning expert knowledge into political, 66

Alexandrova, Anna, 17, 177n4

Allen, Danielle, 234

Allen, Myles, 208, 209, 211, 214

"alphabet soup agencies," 29

Al Roumi, Ohood, 194

Altemeyer, Bob, 55–56

Angell, James, 228–29, 252n8

anticipation, moral economy of, 112

antifoundationalism, 233, 253n12

approval ratings, 42

Arendt, Hannah, 233

Aristotle, 106, 107, 164, 183, 227

Arum, Richard, 77

assessment: versus appraisal in health technology, 114n2; numbers dominate, 4–5; rituals of, 35. *See also* evaluation; learning outcomes assessment (LOA)

Association of American Colleges and Universities, 71

Auden, W. H., 19n3

audit narratives, 71–92; auditor figure constructed in, 82, 85; defined, 72–73, 76–77; discredit self-regulating institutions, 85–89; implicit and explicit, 77; learning and teaching reshaped by, 89; in learning assessment discourse, 76–77; as meta-accounts, 77; organizational actors refigured by, 73, 75; for representing the auditee, 85; rhetoric in, 77, 81, 89; shape how we think about organizational power structures, 78; values and, 73, 75, 77, 85, 87, 88, 91

audits: for accountability, 5, 74, 78; auditee as bad actor and responsible subject, 85–89; auditor as mediator, 81–85; culture of, 72–73, 75, 76, 89, 90, 91, 92n4; objectivity, translatability, and accountability attributed to, 5; as performative, 75, 76, 81, 90–91;

Made in United States
Orlando, FL
27 August 2022

21618418R10176